Grounding and Bonding Photovoltaic and Energy Storage Systems

RIVER PUBLISHERS SERIES IN ENERGY SUSTAINABILITY AND EFFICIENCY

Series Editors:

PEDRAM ASEF
Lecturer (Asst. Prof.) in Automotive Engineering,
University of Hertfordshire, UK

The "River Publishers Series in Sustainability and Efficiency" is a series of comprehensive academic and professional books which focus on theory and applications in sustainable and efficient energy solutions. The books serve as a multi-disciplinary resource linking sustainable energy and society, fulfilling the rapidly growing worldwide interest in energy solutions. All fields of possible sustainable energy solutions and applications are addressed, not only from a technical point of view, but also from economic, social, political, and financial aspects. Books published in the series include research monographs, edited volumes, handbooks and textbooks. They provide professionals, researchers, educators, and advanced students in the field with an invaluable insight into the latest research and developments.

Topics covered in the series include, but are not limited to:

- Sustainable energy development and management;
- Alternate and renewable energies;
- Energy conservation;
- Energy efficiency;
- Carbon reduction;
- Environment.

For a list of other books in this series, visit www.riverpublishers.com

Grounding and Bonding Photovoltaic and Energy Storage Systems

Gregory P. Bierals

Electrical Design Institute, USA

NEW YORK AND LONDON

Published 2025 by River Publishers
River Publishers
Alsbjergvej 10, 9260 Gistrup, Denmark
www.riverpublishers.com

Distributed exclusively by Routledge
605 Third Avenue, New York, NY 10017, USA
4 Park Square, Milton Park, Abingdon, Oxon OX14 4RN

Grounding and Bonding Photovoltaic and Energy Storage Systems /
by Gregory P. Bierals.

© 2025 River Publishers. All rights reserved. No part of this publication may
be reproduced, stored in a retrieval systems, or transmitted in any form or by
any means, mechanical, photocopying, recording or otherwise, without prior
written permission of the publishers.

Routledge is an imprint of the Taylor & Francis Group, an informa
business

ISBN 978-87-7004-145-4 (hardback)
ISBN 978-87-7004-779-1 (paperback)
ISBN 978-87-7004-726-5 (online)
ISBN 978-87-7004-725-8 (master ebook)

While every effort is made to provide dependable information, the
publisher, authors, and editors cannot be held responsible for any errors
or omissions.

Contents

Preface   vii

List of Abbreviations   ix

1   Introduction   1

2   PV Source and Output Circuit Calculations   33

3   Single-Phase Fault Current Analysis   55

4   Three-Phase Fault Current Analysis   77

5   Exercise   117

Index   133

About the Author   135

Preface

This book is designed for engineers, designers, electricians, maintenance personnel, and solar professionals who want to further their understanding of proper grounding and bonding methods for all types of electrical distribution systems, but especially for photovoltaic and energy storage systems.

Chapter 1 includes the terms and definitions that are associated with the topics of grounding electrical systems and equipment, as well as proper bonding methods. This is especially important for PV systems due to their exposure to lightning and external power faults which have the potential to subject the connected electrical load to high frequency and short duration fault currents that are potentially hazardous to people, as well as to equipment.

Chapter 2 includes information on PV source and output circuit calculations and the grounding and bonding methods relating to PV installations.

Chapters 3 and 4 include a complete summary of a single-phase and a three-phase distribution system, beginning at the source and continuing downstream to utilization equipment. All circuit components are sized in accordance with the connected electrical load, including normal load currents, as well as the calculated fault currents.

Chapter 5 includes a 50 question quiz and an answer key which is intended to further your understanding of this important subject.

In summary, this book will serve to provide individuals with an in-depth understanding of how to apply critical grounding and bonding solutions to photovoltaic and energy storage systems.

As always, I welcome your comments and suggestions.

List of Abbreviations

EGC	Equipment Grounding Conductor
EMT	Electrical Metallic Tubing
GEM	Ground-Enhancement Material
GFCI	Ground-Fault Circuit Interrupter
GFP	Ground-Fault Protection
IE	Incident Energy
MCOV	Maximum Continuous Operating Voltage
NEC	National Electrical Code
OCPD	Overcurrent Protective Device
PPE	Personal Protective Equipment
PVC	Polyvinyl Chloride Conduit
RMC	Rigid Metal Conduit
SPD	Surge Protective Device
THHN	90 Degree-C Thermoplastic Insulation

1

Introduction

This chapter identifies the important definitions of the terms associated with grounding systems and equipment. Each defined term is provided with a detailed explanation, along with the appropriate NEC references that may be reviewed in order to further enhance the understanding of these terms.

We begin our analysis of grounding electrical systems and equipment with a detailed analysis of the definitions of the appropriate terms in Article 100. This summary will include related information, including the insulation withstand rating of conductors, the withstand rating of equipment terminations, and the fusing or melting currents of copper conductors. In addition, there are examples of calculations to further enhance the understanding of this important topic and the relationship to proper grounding and bonding methods. The most important aspect of grounding and bonding is simply that the integrity of the conducting paths associated with either of these critical systems be designed to assure that they are properly sized for the calculated ground-fault current and the duration of this current flow in accordance with the operating characteristics of the circuit overcurrent device. Too often, the equipment grounding conductor is sized from 250.122 without consideration of the actual ground-fault current in the circuit. This is a potential hazard and certainly unacceptable, both from the standpoint of people protection, as well as the protection of equipment.

DEFINITIONS

Appliance-

Utilization equipment, generally other than industrial, that is fastened in place, stationary, or portable, and normally built in a standardized size or type and is installed or connected as a unit to perform one or more functions,

2 Introduction

such as clothes washing, air-conditioning, food mixing, deep frying, and so
forth (220.11, 220.14, 220.50, 220.52, 220.53, 220.54, 220.55, 220.56).

Approved-

Acceptable to the authority jurisdiction; usually the electrical inspector.
NFPA 79 amd NFPA 791 are important standards for evaluating equipment
on-site.

Fault Current-

The current delivered at a point on the system during a short-circuit condition.

Fault-Current, Available-

The largest amount of current capable of being delivered at a point on the
system during a short-circuit condition. A fault between circuit conductors
may be caused by connections of ungrounded conductors, ungrounded conductor and grounded conductor, or ungrounded conductors and equipment
grounding conductor.

Fault-Protection Device-

An electronic device that is intended for the protection of personnel and
functions under fault conditions, such as network-powered broadband communications cable short or open circuit, to limit the current or voltage, or
both, for a low-power network-powered broadband communications circuit
and provide acceptable protection from electric shock.

Ground-Fault Current Path, Effective-

This term defines the ground-fault conducting path, from the point where
the ground-fault occurs, to the source of the electrical supply (transformer,
generator, etc.).

This effective ground-fault current path must have three components.
These are:

1) The path must be permanent and continuous (300.10), (250.96(A)),
   (250.97), (250.98), (250.100).
2) The path must have ample capacity to safely carry the ground-fault
   current likely to be imposed on it. This provision shows the importance
   of understanding the short-time ampere ratings of conductors, as the

Introduction 3

ground-fault current may be quite high but typically of a short duration until the overcurrent device operates to clear the ground-fault. The amount of ground-fault current that is expected to flow through this conducting path, and the duration of the ground-fault, or the time that it takes for the overcurrent device to clear this fault, must be known in order to satisfy this requirement (250.4(A)(5)), (250.4(B)(4)), (250.118), (250.122), (250.134(1)(2)).

For example, let us say that the system is solidly grounded and the calculated ground-fault current on a 100-ampere circuit is *13,950 amperes*. This circuit is protected by a 100-ampere, three-pole molded-case circuit breaker and the conductors are 3 THHN copper (26.66 mm^2). At 13,950 amperes, it has been determined that this circuit breaker will clear this ground-fault in 0.025 seconds(60 Hz). The equipment grounding conductor for the 100-ampere circuit is 8 THHN copper (8.37 mm^2), taken from the minimum size equipment grounding conductor Table 250.122, based on the 100-ampere circuit rating. The insulation withstand rating for 0.025 seconds is ask follows:

8 AWG – 16.510 cm (Table 8-Chapter 9) –one ampere for every 42.25 circular mils of conductor circular mil area for 5 seconds

$$\frac{16.510 \text{ cm}}{42.25} = 390.77 \text{ amperes}$$

$$390.77 \text{ A} \times 390.77 \text{ A} \times 5 = 763,505.96$$

$$\frac{763,505.96}{0.025} = 30,540,238.40$$

$$\sqrt{30,540,238.40} = 5526.32 \text{ A}$$

The insulation withstand rating of the 8 AWG copper conductor for 0.025 seconds is 5,526.32 amperes.

However, the calculated ground-fault current in this circuit is 13,950 amperes. If a ground-fault occurred in this circuit, the insulation on this 8 AWG copper wire would be destroyed. This wire would not provide an effective ground-fault current path. Not only would this violate the second provision of this definition, but this would also be a violation of Sections 110.10 and 110.3(B).

4 Introduction

The solution would be to change the type of overcurrent device (faster clearing time), or increase the size of the equipment grounding conductor.

By increasing the conductor size to 4 AWG, the insulation withstand rating is increased to 13,971.44 amperes, and this is acceptable.

4 AWG − 41, 740 cm (Table 8-Chapter 9)

$$\frac{41,740 \text{ cm}}{42.25} = 987.93 \text{ A}$$

93 A × 987.93 A × 5 seconds = 4, 880, 028.42

$$\frac{4,880,028.42}{0.025} = 195,201,137$$

$$\sqrt{195,201,137} = 13,971.44 \text{ A}$$

The equipment grounding conductor for this circuit is 4 AWG copper if the molded-case circuit breaker is used, or 6 AWG copper if a current-limiting overcurrent device is the circuit overcurrent protective device, as the clearing time of this protective device is less than 0.008 seconds (one cycle) (60 Hz). It must be noted that the operating characteristics of the overcurrent device must be determined to assure proper circuit protection.

6 AWG − 26.240 cm

$$\frac{26.240 \text{ cm}}{42.25} = 621 \text{ A} - 5 \text{ seconds}$$

621 A × 621 A × 5 seconds = 1, 928, 205

$$\frac{1,928,205}{0.008} = 241,025,625 \quad \left(0.008 \text{ seconds} = \frac{1}{2} \text{ cycle} \right)$$

$$\sqrt{241,025,625} = 15,525A$$

This example shows why Table 250.122 only identifies the minimum sizes of equipment grounding conductors.

Introduction 5

INSULATION WITHSTAND RATINGS
150 °C MAXIMUM

AWG	Normal 75 °C	5 sec. 150 °C	1 sec. 150 °C	1 cycle - 0.016 sec. 150 °C	1/2 cycle - 0.008 sec. 150 °C	1/4 cycle - 0.004 sec. 150 °C	1/8 cycle - 0.002 sec. 150 °C
14	20 A	97 A	217 A	1,715 A	2,425 A	3,429 A	4,850 A
12	25 A	155 A	347 A	2,740 A	3,875 A	5,480 A	7,750 A
10	35 A	246 A	550 A	4,349 A	6,150 A	8,697 A	12,300 A
8	50 A	397 A	888 A	6,912 A	9,775 A	13,824 A	19.550 A
6	65 A	621 A	1,389 A	10,977 A	15,525 A	21,956 A	31,050 A
4	85 A	988 A	2,209 A	I7,466 A	24,700 A	34,931 A	49,400 A
3	100 A	1,245 A	2,784 A	22,008 A	31,125 A	44,017 A	63,450 A
2	115 A	1,571 A	3,513 A	27,772 A	39,275 A	55,543 A	78,550 A
1	130 A	1,981 A	4,430 A	35,019 A	49,525 A	70,039 A	99,050 A
1/0	150 A	2,499 A	5,588 A	44,176 A	62,475 A	88,353 A	124,950 A
2/0	175 A	3,150 A	7,044 A	55,685 A	78,750 A	111,369 A	157,500 A
3/0	200 A	3,972 A	8,882 A	70,216 A	99,300 A	140,431 A	198,600 A
4/0	230 A	5,008 A	11,198 A	88,530 A	125,200 A	177,060 A	250,400 A
250 kcmi	25SA	5,917 A	13,231 A	104,599 A	147,925 A	209,198 A	295,850 A
300 kcmi	285 A	7,101 A	15,878 A	125,529 A	177,525 A	251,058 A	355,050 A
350 kcmi	310 A	8,284 A	18,524 A	146,442 A	207,100 A	292,884 A	414,200 A
400 kcmi	335 A	9,467 A	21,169 A	I67,354 A	236,675 A	334,709 A	473,350 A
500 kcmi	380 A	11,834 A	26,462 A	209,198 A	295,850 A	418,395 A	591,700 A

Example

Insulation Withstand Rating

No. 6 AWG copper – 65 amperes (continuous) $-75\,°C$

No. 6 AWG copper $-26,240$ circular mils -13.30 mm^2

I^2T – ampere – squared seconds

I^2T – one ampere for every 42.25 circular mils of conductor cross-sectional area for 5 seconds

No. 6 AWG $-\dfrac{26,240 \text{ circular mils}}{42.25} = 621$ amperes -5 seconds

To determine the insulation withstand rating for one cycle (0.016 seconds), the calculation is as follows:

621 amperes $\times 621$ amperes $\times 5$ seconds $= 1,928,205$

$\dfrac{1,928,205}{.016} = 120,512,813$

$\sqrt{120,512,813} = 10,978$ amperes

6 *Introduction*

Therefore, the one-cycle insulation withstand rating is 10,978 amperes, which will produce a temperature of 150°C in this conductor. This is the maximum temperature that the insulation can safely withstand without damage.

TERMINAL WITHSTAND RATINGS
250°C MAXIMUM

AWG	Normal 75 °C	5 sec. 250 °C	1sec. 250 °C	1 cycle - 0.016 sec. 250 °C	1/2 cycle - 0.008 sec. 250 °C	1/4 cycle - 0.004 sec. 250 °C	1/8 cycle - 0.002sec. 250 °C
14	20 A	112 A	250 A	1,980 A	2,800 A	3,960 A	5,600 A
12	25 A	178 A	398 A	3,147 A	4,450 A	6,293 A	8,900 A
10	35 A	282 A	631 A	4,985 A	7,050 A	9,970 A	14,100 A
S	50 A	449 A	1,004 A	7,937 A	11,225 A	15,875 A	22,450 A
6	65 A	714 A	1,597 A	12,622 A	17,850 A	25,244 A	35,700 A
4	85 A	1,136 A	2,540 A	20,082 A	28,400 A	40,164 A	56,800 A
3	100 A	1,459 A	3,262 A	25,792 A	36,475 A	51,583 A	72,950 A
2	115 A	1,806 A	4,038 A	31,926 A	45,150 A	63,852 A	90,300 A
1	130 A	2,277 A	5,092 A	40,252 A	56,925 A	80,504 A	113,850 A
1/0	150 A	2,873 A	6,424 A	50,788 A	71,825 A	101,576 A	143,650 A
2/0	175 A	3,622 A	8,099 A	64,029 A	90,550 A	128,057 A	181,100 A
3/0	200 A	4,566 A	10,210 A	80,716 A	114,150 A	161,432 A	228,300 A
4/0	230 A	5,758 A	12,875 A	101,788 A	143,950 A	203,576 A	287,900 A
250 kcmil	255 A	6,803 A	15,212 A	120,261 A	170,075 A	240,522 A	340,150 A
300 kcmil	285 A	8,163 A	18,253 A	144,303 A	204,075 A	288,606 A	408,150 A
350 kcmi	310 A	9,524 A	21,296 A	168,362 A	238,100 A	336,724 A	476,200 A
400 kcmi	335 A	10,884 A	24,337 A	192,404 A	272,100 A	384,808 A	544,200 A
500 kcmi	380 A	13,605 A	30,422 A	240,505 A	340,125 A	481,009 A	680,250 A

Example

Terminal Withstand Rating

No. 6 AWG copper – 65 amperes (continuous) $-75\,°C$

No. 6 AWG copper $-26,240$ circular mils -13.30 mm^2

I^2T-1 ampere for every 36.75 circular mils of conductor cross-sectional area for 5 seconds

No. 6 AWG $-\dfrac{26,240 \text{ circular mils}}{36.75} = 714$ amperes –5 seconds

To determine the conductor terminal withstand rating for one cycle (0.016 seconds), the calculation is as follows:

714 amperes $\times 714$ amperes $\times 5$ seconds $= 2,548,980$

$$\frac{2,548,980}{.016} = 159,311,250$$

$$\sqrt{159,311,250} = 12,622 \text{ amperes}$$

Therefore, the one-cycle (0.016 seconds) terminal withstand rating is 12,622 amperes, which will produce a temperature of 250 °C. This is the maximum temperature that the terminal can safely withstand. That is, after reaching this temperature and cooling after the fault is cleared, the integrity of the terminal is still intact.

FUSING OR MELTING CURRENT
$1,083°$ C MAXIMUM

	Normal	5 sec.	1 sec.	1 cycle - 0.016 sec.	1/2 cycle - 0.008 sec.	1/4 cycle - 0.004 sec.	1/8 cycle - 0.002 sec.
AWG	75 °C	1,083 °C	1,083 °C	1,083 °C	1,083 °C	1,083 °C	1,083 °C
14	20 A	254 A	568 A	4,490 A	6,350 A	8,980 A	12,700 A
12	25 A	403 A	901 A	7,124 A	10,075 A	14,248 A	20,150 A
10	35 A	641 A	1,433 A	11,331 A	16,025 A	22,663 A	32,050 A
8	50 A	1,020 A	2,281 A	18,031 A	25,500 A	36,062 A	51,000 A
6	65 A	1,621 A	3,625 A	28,656 A	40,525 A	57,311 A	81,050 A
4	85 A	2,578 A	5,765 A	45,573 A	64,450 A	91,146 A	128,900 A
3	100 A	3,312 A	7,406 A	58,548 A	82,800 A	117,097 A	165,000 A
2	15 A	4,101 A	9,170 A	72,461 A	102,475 A	144,922 A	204,950 A
1	130 A	5,169 A	11,558 A	91,376 A	129,225 A	183,105 A	258,950 A
1/0	150 A	6,523 A	14,586 A	115,311 A	163,075 A	230,623 A	326,150 A
2/0	175 A	8,221 A	18,383 A	145,328 A	205,525 A	290,656 A	411,050 A
3/0	200 A	10,364 A	23,175 A	183,211 A	259,100 A	366,423 A	518,200 A
4/0	230 A	13,070 A	29,225 A	231,047 A	326,750 A	462,094 A	653,500 A
250 kcmil	255 A	15,442 A	34,529 A	272,979 A	386,050 A	545,957 A	772,100 A
300 kcmi	285 A	18,530 A	41,434 A	327,567 A	463,250 A	655,134 A	925,500 A
350 kcmil	310 A	21,618 A	48,339 A	382,156 A	540,450 A	764,312 A	1,080,900 A
400 kcmi	335 A	24,707 A	55,247 A	436,762 A	617,675 A	873,524 A	1,235,350 A
500 kcmil	380 A	30,883 A	69,056 A	545,939 A	772,075 A	1,091,879 A	1,544,150 A

Example

Fusing or Melting Current

No. 6 AWG copper – 65 amperes (continuous) – 75 °C

No. 6 AWG copper – 26,240 circular mils – 13.30 mm^2

I^2T – 1 ampere for every 16.19 circular mils of conductor cross-sectional area for 5 seconds

I^2T – ampere – squared seconds

8 *Introduction*

$$\text{No. 6 AWG} - \frac{26,240 \text{ circular mils}}{16.19} = 1621 \text{ amperes} - 5 \text{ seconds}$$

To calculate the fusing or melting current for one cycle (0.016 seconds), the calculation is as follows:

$$1{,}621 \text{ amperes} \times 1{,}621 \text{ amperes} \times 5 \text{ seconds} = 13{,}138{,}205$$

$$\frac{13,138,205}{.016} = 821,137,813$$

$$\sqrt{821,137,813} = 28,656 \text{ amperes}$$

Therefore, the one-cycle (0.016 seconds) fusing or melting current of this conductor is 28,656 amperes, which will produce a temperature of 1,083°C, after which this the conductor will begin to fuse or melt.

AWG - TO METRIC CONVERSION CHART

AWG	CIRCULAR MII AREA	METRIC SIZE - MM^2
14	4110 cm	2.08
12	6530 cm	3.31
10	10,300 cm	5.261
8	16,510 cm	8.367
6	26,240 cm	13.3
4	41,740 cm	21.15
3	53,620 cm	26.67
2	66,360 cm	33.62
1	83,690 cm	42.41
1/0	105,600 cm	53.49
2/0	133,100 cm	67.43
3/0	167,800 cm	85.01
4/0	211,600 cm	107.2
250 Kcmil	250,000 cm	127
300 Kcmil	300,000 cm	152
350 Kcmil	350,000 cm	177
400 Kcmil	400,000 cm	203
500 Kcmil	500,000 cm	253

$$\text{e.g. - No. 2 AWG} = \frac{66,360 \text{ Cricular Mils}}{1973.53} = 33.625 \text{ mm}^2$$

I would like to point out that if the fusing current of the 8 AWG copper wire was determined from the chart in this book, the fusing current for one

Introduction 9

cycle is referenced at 18,031 amperes. Once again, if our calculated ground-fault current is 13,950 amperes, then the minimum size equipment grounding conductor, 8 AWG copper from Table 250.122, would be acceptable.

However, my suggestion is to use the chart based on insulation withstand ratings. This is because the equipment grounding conductor is within the same raceway or cable with the other circuit conductors (for important impedance reduction) (Section 300.3(B)). And if it were an insulated wire, this would serve to protect the other conductors from the thermal-stress of the ground-fault. It should be noted that 250.134, Exception No. 2, permits the equipment grounding conductor for DC circuits to be run separately from the circuit conductors.

However, 690.43(C) requires the equipment grounding conductors for the PV array and support structure to be contained within the same raceway or cable or otherwise run with the PV system conductors where these conductors leave the vicinity of the PV array, even though these are DC circuits. So, 690.43(C) amends 300.3(B) and this is in accordance with 90.3.

Another possible solution would be to use a current-limiting circuit breaker instead of the standard device. Current-limiting devices (fuses or circuit breakers) operate in less than 1/2 cycles (0.008 seconds) when they are interrupting currents that are within their current limiting range (short-circuit currents reach a peak in the first 1/2 cycle). So, instead of a 0.025-second clearing time for the standard circuit breaker, we may have reduced the fault-clearing time to less than 1/2 cycle (0.008 seconds), and this would increase the insulation withstand rating for the AWG copper wire (8.37 mm^2) to at least 9,775 amperes. However, the available ground-fault current in our example is 13,950 amperes. So, the 8 AWG copper wire is not acceptable as an equipment grounding conductor for this circuit. Using a 6 AWG insulated copper wire (13.30 mm^2) for the equipment grounding conductor would be acceptable if the current-limiting overcurrent device is used or 4 AWG insulated copper if the molded-case circuit breaker is used.

This is a good example of using the conductor short-time rating charts in this book to determine the validity of the effective ground-fault current path.

3) The ground-fault current path must be of sufficiently low impedance to limit the voltage-to-ground and facilitate the operation of the over-current device (on a solidly-grounded system). During a ground-fault, as the equipment grounding conductor is carrying the ground-fault current back to the source (transformer, generator, etc.), the voltage-drop associated with this current flow produces a voltage-rise, above earth

10 *Introduction*

potential (0 volts), on anything connected to the equipment grounding conductor. This may produce an excessive voltage-rise on equipment. Possibly a dangerous touch-potential may appear on equipment metal frames. So, the equipment grounding conductor must be properly sized to limit the voltage-drop and resultant voltage-rise. Where a change occurs in the size of the ungrounded conductors to compensate for voltage-drop, a proportional increase must be made in the size of the equipment grounding conductor (250.122(B)). This provision may not be applicable for dc circuits, unless they are excessively long. For example, 690.45 states that increasing the size of the EGC to compensate for voltage-drop for solar photovoltaic source and output circuits is not required (250.4(A)(2)(3)(4)(5)), (250.4(B)(2)(3)(4)).

Voltage-Drop Example

Branch-circuit rating – 40 Amperes

Circuit voltage – 240 volts – Single-phase circuit

Load 32 amperes (continuous)

Normal ungrounded conductor size – 8 AWG copper

Minimum equipment grounding conductor size – No. 10 AWG copper (Table 250.122)

Circuit length (from source to load) – 270 feet – (82.3 meters)

According to Section 210.19. Informational Note, voltage-drop for a combination feeder and branch-circuit should be limited to 5% of the applied voltage to provide for reasonable efficiency of operation. For a branch-circuit, the voltage-drop should be limited to 3%. For a feeder, this Informational Note expresses a voltage-drop of 3% of the applied voltage as the recommended limit for reasonable efficiency of operation.

This information is repeated in Section 215.2(A)(2), Informational Note No. 2.

240 volts

$$\frac{2K \times L \times I}{cm} \times 0.03(210.19, \text{Informational Note})$$
7.2 volts

Where K = 12.9 - Copper
L = One way length in feet of conductors
I = Amperes of load

Introduction 11

Equipment grounding conductor–40 ampere circuit–10 AWG copper (Table 250.122)

$$\frac{25.8 \times 270 \text{ feet} \times 32 \text{ amperes}}{10,380 \text{ circular mils}(10\text{AWG})} = \frac{222,912}{10,380 \text{ cm}} = 21.48 \ V$$

$$\frac{25.8 \times 270 \text{ feet} \times 32 \text{ amperes}}{26,240 \text{ circular mils}(6 \text{ AWG})} = \frac{222,912}{26,240 \text{ cm}} = 8.50 \ V$$

$$\frac{25.8 \times 270 \text{ feet} \times 32 \text{ amperes}}{41,740 \text{ circular milsl} (4 \text{ AWG})} = \frac{222,912}{41,740 \text{ cm}} = 5.34 \ V$$

Equipment grounding conductor size – 4 AWG – Copper (250.122 (B))

Remember, as Informational Notes, this information is not enforceable (90.5(C)). So lower or even higher levels of voltage-drop may be acceptable, or even necessary, depending on the type of equipment supplied. Also, 647.4(D) applies to sensitive electronic equipment, where the voltage-drop for this equipment is limited to 1.5% for branch-circuits and 2.5% for a combination feeder and branch-circuit.

- **Ground:** The earth (or to some conducting body that serves in place of the earth).
- As we will discuss in detail later, certain conducting bodies may serve as a ground reference. For example, the metal frame of a vehicle or the metal and metallic skin and frame of an airplane, and the metal frame of a portable, vehicle-mounted, and trailer-mounted generator (250.34(A)).
- **Ground-Fault:** This is an accidental connection between an ungrounded conductor and the equipment grounding conductor, which may be a wire, metal raceway, metal cable tray, metallic equipment frame, or the earth (250.118).

Based on extensive research, it is safe to conclude that 90% of faults occurring in electrical systems are ground-faults. And, further, 90% of ground-faults are arcing-type faults. In addition, there is a voltage-drop associated with this arcing fault, which has the effect of reducing the amount of ground-fault current in the circuit. And couple this problem with the impedance of the ground-fault return path, and it is easy to see why the circuit overcurrent device may not operate in a short enough period of time, or, at all, to prevent a hazard to people and/or to equipment. And because the ungrounded conductors come in contact with the equipment grounding system in many places throughout a typical distribution system, it is easy to see why a ground-fault is the most common type of fault. It may not be caused by damage or

12   *Introduction*

the age of the insulation but simply by contaminants that have been absorbed through the insulation, such as dirt combined with moisture.

Ground-Fault Circuit Interrupter, Special Purpose (SPGFCI)- A device intended for the detection of ground-fault currents, and used in circuits with voltages to ground greater than 150 volts, that functions to de-energize a circuit or portion of a circuit within an established period of time when a ground-fault current exceeds the values established for Class C, D, or E devices (See UL 943C). This device is designed to recognize ground-fault currents on 480/277V circuits where the fault reaches 20ma (UL943C).

- **Grounded-** This is a connection to ground or to a conducting body that extends the ground connection, such as an interior grounded metal water pipe or grounded metal raceways.
- **Grounded, Solidly-** This is a physical connection to ground (earth) through a connection without a resistance element or impedance device in series with this connection (250.20(A)(B)(C)).

I used to ask a question in my grounding classes whether the equipment grounding conductor (metal raceway, metal cable tray, or wire) would carry any current during normal conditions. And the answer to this question was an emphatic "no." But, the actual answer is "yes." The equipment grounding conductor is installed within the same raceway, cable tray, cable, or cord as the other circuit conductors (ac systems). Or, it is the metallic raceway or cable tray itself (300.3(B)), (300.5(1))), (392.60(A)), (392.60(B)), (250.96)), (250.102). Therefore, some of the current flowing in the circuit conductors is magnetically and capacitively coupled into the equipment grounding conductor. This is known as an inductive- and capacitive-charging current, and even on large systems, it should be quite low.

If higher current is present in the equipment grounding conductor, it may be indicative of upstream cross-bonds between a grounded (neutral) conductor and the equipment grounding conductor. Or, as I have seen in some installations, the grounded neutral conductor is used as a means of grounding equipment enclosures. It should be noted that at least up to the 1993 NEC, the neutral conductor was permitted to be used to ground the equipment frames of electric ranges and cloth dryers (250.140). This was typically done through the use of Type SE cable, which has an uninsulated neutral conductor as part of the cable assembly. This practice was prohibited in the 1996 NEC, with an exception retained for existing installations ((250.140(B)), (250.142(B)), Exception No. 1). Also, for grounding meter enclosures on the load side of

Introduction 13

the service disconnect, the grounded circuit conductor (neutral) may be used for this purpose (250.142(B), Exception No. 2).

- **Grounded Conductor-** This is a conductor that has been intentionally grounded. This conductor is, most often, referred to as the "neutral" conductor. On single-phase, three-wire systems, and three-phase, four-wire, Wye connected systems, the system *neutral point* is solidly connected to ground (earth). The neutral point of these systems is the physical point where voltage from the other points of the system would be equal. For example, a three-phase, Wye-connected transformer secondary, where voltage from X1−X2−X3 to the X0 (neutral point) would be equal. The X0 neutral point may be solidly grounded, and a grounded (neutral) conductor would originate from X0 ((250.24), (250.26)).

However, on a three-phase, delta, corner-grounded system, one phase conductor is intentionally grounded. This conductor is a grounded conductor but not a "neutral" conductor. However, the means of identifying grounded conductors (200.6) applies to neutral conductors, as well as the grounded phase conductor of a corner-grounded delta system.

The grounded conductor is connected to ground in accordance with 250.24(A)(1) through (A)(4). A grounding electrode conductor connection is normally made at any accessible point from the load end of the overhead service conductors, service drop, underground service conductors, or service lateral to, and including the terminal or bus to which the grounded service conductor is connected at the service disconnecting means (250.24(A)(1)). Where the supply transformer is located outside of the building, an additional grounding connection from the grounded service conductor to a grounding electrode is made at the transformer or elsewhere outside the building (250.24(A)(2)). This connection to ground will serve to protect the supply system from lightning-induced currents and other external power faults.

For separately derived systems that are grounded systems, the grounded conductor is connected to the grounding electrode (system) at the source or at the first disconnect or overcurrent device that is supplied from the separately derived system (250.30)(A)(1) through (A)(5) through a Supply-Side Bonding conductor.

In order to prevent objectionable current on the equipment grounding system (250.6(A)) beyond the initial connection between the grounded conductor and the equipment grounding system at or before the service equipment, at the source of the separately derived system, or at the first

14 *Introduction*

disconnect or overcurrent device supplied by the separately derived system, additional connections beyond the initial connection are not permitted.

The equipment grounding system should not be carrying current from the grounded conductor(s), other than the negligible inductive and capacitive charging currents from the other conductors that are run with the equipment grounding conductors.

The current produced by the downstream cross-bonding of these systems will cause a voltage-rise on the equipment grounding system that may prove to be a touch-potential problem for people and also damaging to equipment.

In addition, these connections produce grounding loops that may be a problem for certain types of electronic equipment.

Section 200.4 applies to the installation and identification of neutral (or grounded) conductors where multiple circuits are installed in a common enclosure.

Sections 200.6 and 200.7 specify the requirements for identifying grounded conductors, typically by the use of white or gray insulation or by white or gray markings at terminations.

- **Ground-Fault Circuit Interrupter-**

This device was invented by Professor Charles Dalziel at the University of California, (Berkeley), and he received a patent in 1961. A commercially available GFCI (Class B) was introduced in 1964, specifically for wet niche swimming pool lighting fixtures, where the leakage current could have been as high as 20 ma. These GFCIs have not been listed for many years.

The first reference for GFCIs in the NEC appeared in 1968. This device has a differential transformer that surrounds the supply and return conductors and monitors the current difference in these conductors. If the current difference reaches a level of 4—6 ma or higher (Class A GFCI), the unit trips in about 1/40 second. The listing standard for this device is UL 943.

The required use of these devices has expanded significantly over time. One of the original code references, which addresses the use of GFCI's in dwelling and non-dwelling occupancies is Section 210.8 (A)(B).

- **Ground-Fault Current Path-**

This is defined as a conductive path, from the point on the wiring system where the ground-fault develops, and then through any conducting paths back to the source of the electrical supply. Hopefully, the lowest impedance path for this current to flow is through the equipment grounding conductor (250.118). However, other conducting paths may parallel the equipment

Introduction 15

grounding conductor. A portion of the ground-fault current will flow through these parallel paths. Also, when the ground-fault current reaches the service equipment (or ahead of the service equipment), the current will divide, unevenly, on its path back to the source (solidly grounded system). Due to the fact that there will be a grounding electrode connected to the grounded conductor and the equipment grounding system within the service equipment, or ahead of the service equipment, and a grounding electrode connected to the grounded point of the supply transformer, another parallel path is established through the earth between these two grounding electrodes. Some of the normal neutral current will return to the transformer through this earth path, as well as some of the ground-fault current. This is a conducting path, albeit a high impedance path, as compared to the path through the service grounded (neutral) conductor. This will apply whether the supply transformer is owned by the serving utility or by the customer (250.24(A)(2)). It would be safe to say that only a small percentage of neutral current, or ground-fault current, will return to the source through the earth.

- **Ground-Fault Protection of Equipment-**

This protection is a critical part of a solidly grounded three-phase, four-wire, Wye-connected system, which operates at over 150 volts-to-ground, and not in excess of 1,000 volts phase-to-phase where the service disconnect is rated 1000 amperes or more. The most common type of system for this protection is a three-phase, four-wire, solidly grounded, Wye-connected electrical system with a voltage of 480 volts, phase-to-phase, and 277 volts, phase-to-neutral, and having a service disconnect rated at 1,000 amperes or more. This is a type of distribution system where arcing ground-faults may be the most destructive.

It is very important that the pickup setting of the GFP is selectively coordinated with downstream overcurrent devices in order to prevent a nonorderly shutdown of an entire system. Sections 215.10 (Feeders), 230.95 (Services), and 517.17 (Health Care Facilities) address the requirements for this protection. Section 517.17 requires a second level of ground-fault protection on every feeder downstream of this protection at the service equipment. This downstream GFP will have a lower pickup setting than the main GFP to assure that the feeder overcurrent device will open ground-faults on its load side, without affecting the operation of the service overcurrent device. This will limit an outage to a single feeder, so that the rest of the distribution system will remain operational. In addition, Sections 700.6(D) and 701.6(D) require that a *ground-fault sensor* be provided for an emergency system or

16 *Introduction*

a legally required standby system on solidly grounded Wye systems of more than 150 volts-to-ground and with overcurrent devices rated 1,000 amperes, or more. This permits the ground-fault to be detected without interrupting the emergency system or the legally required standby system and the fault cleared when the normal service is restored. For critical operations power systems (708.52(B)), an additional step of ground-fault protection at the next level of feeder disconnecting means downstream toward the load is required.

Types of GFPE

Ground Strap-Type System- This type of protection system incorporates a current sensor, control power source and ground-fault relay along with a fusible disconnect switch or a circuit breaker. The ground-fault sensor surrounds the main bonding jumper or the system bonding jumper for a separately derived system. This type of protection is installed at the source, the service equipment, the source of a separately derived system, or at the first disconnect or overcurrent device supplied by the separately derived system (250.24(C)), (250.28(A),(C),(D)). The disadvantage of this type of ground-fault protection is that it may only be installed at the source or at the first disconnecting means of a separately derived system and not at downstream feeders, as the ground-fault sensor is installed at the location of the main bonding jumper or system bonding jumper.

Zero-Sequence Ground-Fault- Sensing system consists of a current sensor that surrounds all of the current-carrying conductors of the system, including the grounded (neutral) conductor. This type of protection system includes a control power source, a ground-fault relay and a shunt-trip circuit breaker or shunt-trip fuse disconnect. The ground-fault sensor is installed downstream of the main bonding or system bonding jumpers so that these bonding jumpers and the equipment grounding conductor do not pass through the ground-fault sensor. During normal operation, the current sensor should register at or near zero amperes. During a ground-fault, not all of the current from the source returns through the phase and neutral conductors and the sensor detects this current imbalance and signals the ground-fault relay to open the circuit breaker or disconnect switch.

The Residual-Type of Ground-Fault Sensor- Includes the phase sensors built into the circuit breaker. An external ground-fault current sensor will surround the neutral conductor and this sensor may, at times, be field installed.

Introduction   17

This type of protection functions the same as the zero-sequence GFP, in that the vector sum of the phase and neutral currents should be at or near zero amperes. The ground-fault relays or circuit breaker settings are set in the field and they range from 4-1200 amperes. The time-delay settings may range from 0.025 seconds, 1.5 cycles (60 Hz) to one second.

- **Grounding Conductor, Equipment-**

This is the conductor, which may be in the form of a wire, metal raceway, metallic cable assembly, or metal cable tray. Section 250.118(A)(1)-(14) identifies the various types of equipment grounding conductors. These conducting paths must be able to carry the maximum ground-fault current that may be imposed on them. If the equipment grounding conductor is in the form of a wire, Table 250.122 specifies its minimum size, in conjunction with Section 250.4(A)(5) and 250.4(B)(4). Section 250.119(A) covers the identification of an insulated equipment grounding conductor. It is to have a continuous outer finish that is green or green with one or more yellow stripes. However, where conductors are 4 AWG (21.15 mm^2) or larger, other means of identification are acceptable, such as the use of green tape at terminations and other places where the wire is accessible, except in fittings, such as conduit bodies that contain no splices or unused openings (250.119(B)).

Sections 300.3(B) and 300.5(I) require that all of the conductors of the same circuit are to be installed together, and in close proximity to each other to maximize the effects of capacitive and inductive coupling and reduce the overall circuit impedance (ac systems). Section 690.43(C) requires the equipment grounding conductors to be run with PV array circuit conductors where these conductors leave the vicinity of the PV array. Section 250.134, Exception No. 2 permits the equipment grounding conductor for DC circuits to be run separately from the circuit conductors, as only conductor resistance will affect the validity of the grounding path.

Magnetic flux density between conductors decreases with the square of the distance between conductors. So, it is important to keep the spacing between conductors to a minimum for important impedance reduction.

- **Grounding Electrode-**

This is the object that establishes a means of making a connection to the earth.

Section 250.52 identifies the various types of grounding electrodes that are used to make this earth connection.

It is very important that the grounding electrode or grounding electrode system selected for a particular installation be of a type that will assure a low

18 *Introduction*

resistance-to-ground. This concept is not only relative to the safety of people but also to assure the proper operation of equipment. This is especially true, due to the sensitivity of the sophisticated electronic equipment in use today. And this may be more difficult now, as this equipment may be installed in an existing installation where a proper grounding system was not considered a priority.

In the past several years, I have had the privilege of working in many countries throughout Africa, as well as Haiti, and even in several locations in North Korea. The installations were in healthcare facilities. In the vast majority of these installations, a proper grounding system was not available, and, in some instances, there was no grounding system at all.

As an example, I was called upon to check the grounding system at the oldest government hospital in Addis Ababa, Ethiopia. GE Medical Systems was providing a new CT scanner for this hospital. And it has been my experience in other installations that GE requires a grounding electrode system to have a resistance-to-ground of less than 2 ohms. Depending on the type of soil, as well as other conditions, this low resistance-to-ground requirement can be difficult to achieve. I solved this problem at another hospital in Ethiopia through the use of a ground ring, supplemented by several driven ground rods. But, that was a new installation and there were no area restrictions. This government hospital in Addis Ababa was in a confined area, so a ground ring was not an option. My preferred grounding electrode would have been a concrete-encased electrode. But this was an existing building, so there was no access to the rebar in the concrete footings without disturbing the concrete. And 250.50, Exception does not require the concrete-encased electrodes of existing buildings or structures to be a part of the grounding electrode system.

Another option that I had used at other installations was a chemically charged ground rod, which may have produced the desired low resistance-to-ground. But, this type of rod (UL 467J) was not available in Ethiopia.

So, I decided to install a group of driven ground rods, 10 feet (3.048 meters) long × 5/8 inch (15.875 mm) in diameter. These were copper-clad steel rods, and they were properly spaced (a minimum of 2 rod lengths, or 20 feet (6.096 meters)) (see the Informational Note following Section 250.53(A)(3)). In order to achieve a resistance-to-ground of less than 2 ohms, 13 paralleled ground rods were installed.

This may sound unrealistic, but before the installation of a driven or buried electrode, it is critical to perform a soil-resistivity test in order to determine the best location for the buried electrode or system.

Introduction 19

There are three conditions that affect the resistance-to-ground of a driven or buried grounding electrode.

1) The metallic mass of the electrode. This portion of the resistance-to-ground is negligible. For example, the electrical resistance of a typical ground rod is quite low. A 5/8-inch diameter copper-clad steel ground rod has the approximate current-carrying capacity of a 3/0 AWG (85 mm^2) copper conductor (200 amperes at 75°C) and a DC resistance of 0.0766 ohms per 1,000 feet (uncoated copper) (75°C) (Table 8-Chapter 9). A buried metal plate (250.52(A)(5) exposes not less than two square feet (0.186m2) of surface to the soil (250.52(A)(7). This area of exposure is significantly more than a typical ground rod. However, 250.53(A)(5) permits the metal plate to be 30 inches (750mm) below the surface of the earth and the soil resistivity at this depth may be too high for the metal plate to have a low resistance-to-ground. Of course, burying the metal plate deeper may produce a lower resistance-to-ground.

2) The resistance of the metal/soil interface. This is also quite low, unless there is a nonconductive coating on the metal mass of the electrode.

3) The resistivity of the soil surrounding the electrode. This is the component that will determine the resistance-to-ground of the grounding electrode. And soil resistivity varies from place to place, and in some instances, even in the same area.

It should be noted that highly conductive soil is also corrosive, which will have an effect on the service life of the electrode.

It would certainly be beneficial, and yes, very often critical, to locate an area where the soil resistivity is the lowest of the surrounding area, as that is the best location for the electrode. This value is expressed in ohm-centimeters. An ohm-centimeter is the electrical resistance across the faces of a cubic centimeter of soil. Where the soil resistivity is high, or unknown, soil enhancements are available. Bentonite (natural clay) has been used for this purpose. This material contains aluminum, iron, magnesium, and sodium. When this material comes in contact with water, it expands. Buried conductors may be surrounded with this material. Just like the typical soil surrounding a driven ground rod, the material within 6 inches (15.24 centimeters) of the rod or wire will have a marked effect on the total resistance-to-ground of the rod or conductor.

Erico "GEM" (ground-enhancement material) is even a better material, as its resistivity is lower than Bentonite (Erico.com) (1-800-248-weld). In addition, this material is noncorrosive and it adheres to the rod or conductor more effectively than Bentonite. Its resistivity is about 10% of Bentonite.

20   *Introduction*

Section 250.52(A) lists the acceptable grounding electrodes. They are:

1) A metal underground water pipe. In order to qualify as a suitable grounding electrode, there must be at least 10 feet (3.048 meters) of metallic underground water pipe in direct contact with the earth. Since the 1978 NEC, underground metal water pipes must be supplemented by at least one additional grounding electrode. This is, very often, a ground rod(s). The reasoning here is that, even though the metal underground water pipe may be extensive, and, as such, the resistance-to-ground of this pipe may be quite low, if repairs to this pipe become necessary, some metallic portions of the pipe may be replaced with nonmetallic pipe.
But, keep in mind that if a ground rod is used as the supplemental electrode, its resistance-to-ground must not exceed 25 ohms (250.53(A)(2), Exception). If the resistance-to-ground exceeds 25 ohms, the single rod must be supplemented by an additional electrode, which may be an additional ground rod.

2) The metal frame of a building or structure that has been effectively grounded. This includes at least one *vertical* metal member of the building that is in direct contact with the earth for 10 feet (3.048 meters) or more. This structural metal member may, or may not be encased in concrete. Or, the tie-down bolts that secure a structural metal member (steel column) to a concrete footing, with the tie-down bolts connected to the reinforcing rods in the footing. Also, the metal frames of buildings that are connected to the concrete-encased electrode are permitted to serve as the grounding electrode conductor, that is, the metal structural frame is permitted to be used as a conductor to interconnect grounding electrodes that are a part of the grounding electrode system, or as a grounding electrode conductor (250.68(C)(2).

3) A concrete-encased electrode. (sometimes referred to as an Ufer ground). This grounding electrode has been in use since 1942, and typically there is a very large amount of reinforcing rod, encased in concrete, in direct contact with the earth. As long as the concrete is not separated from the earth by a vapor barrier, or another type of nonconductive coating, this grounding electrode has a very low resistance-to-ground.

4) Ground Ring. This is the grounding electrode that I used for the new Radiology building at Soddo Christian Hospital in Ethiopia. A new digital X-ray machine and a new CT scanner were to be installed, as well as two UPS units, and possibly an MRI sometime later. Once again, GE required a grounding electrode system with a resistance-to-ground

Introduction 21

of less than 2 ohms. The concrete footings were already poured, and so I decided to use a ground ring. The ring consisted of a 2 AWG (33.63 mm^2) solid copper wire, buried to a depth of 3 feet (1 meter). The ground ring was supplemented by 8—5/8" × 10' copper-clad steel ground rods. The length of the ground ring conductor was 350 feet (107 meters). A three-phase, 380/220-volt, 165-kVa generator was installed at a later date, and the generator neutral point was bonded to this ground ring. So, everything in that building is at, virtually, the same potential so that there is no ground potential difference between the service supply and the generator, as well as the other electrical equipment in this building. The ground ring conductor may serve as an important means of bonding separate system grounding electrodes to form one common grounding electrode system (250.50, 250.58, 250.60, 250.106, 800.100, 805.93(A)(B), 810.21(J), 820.100, 830.93). This also includes buildings where there may be separate electric services due to a large square foot area, where the bonding of the grounding electrodes of these systems may not be required, but this bonding would be beneficial.

5) Rod and Pipe Electrodes. Ground rods are very common, but, they may be very ineffective. They are tested in accordance with UL 467. A copper-clad steel rod has an average service life of 30 years. This service life is dependent upon soil conditions, in that, if the soil has a low resistivity, let us say, less than 10,000 ohm-centimeters, it is considered highly corrosive. From 10,000 to 30,000 ohm-centimeters, the soil is considered mildly corrosive. And above 30,000 ohm-centimeters, it is considered noncorrosive.

The sphere of influence of a ground rod is established by its depth. That is, a ground rod driven to a depth of 10 feet (3.0 m) has a sphere of influence that extends radially 10 feet (approximately) from the ground rod and 10 feet beneath the rod. No other grounding electrodes should occupy this sphere it. Therefore, where ground rods are installed in parallel, for optimum efficiency, they should be separated by twice their driven depths. With this spacing, the resistance-to-ground of the paralleled ground rods will be approximately 50% of the single ground rod (250.53(A)(3), Informational Note) (NFPA 780).

Transmission towers and wind turbine towers typically have galvanized steel and concrete foundations that normally provide an effective grounding electrode system (250.52(A)(3)). The decision to supplement the rebar with a driven electrode(s) is strictly up to the designer or installer. However, the use of a galvanized steel ground rod(s) or a stainless steel ground rod(s) in close proximity to the galvanized steel foundation is certainly a better option, due

22 Introduction

to corrosion caused by galvanic action. This also applies to locations where there are buried steel piping systems or underground steel tanks. 694.40(B)(1) requires a wind turbine tower to be connected to a grounding electrode (system). This grounding electrode (system) is required to be bonded to the premises grounding system (694.40(B)(2), (250.50), (250.58)). Guy wires used for the support of turbine towers are not required to be connected to an equipment grounding conductor or to comply with 250.110. However, in areas that are subject to greater thunderstorm activity, the grounding of the guy wires may be required (694.40(B)(4)).

It should be noted that soils that are highly conductive (soil resistivity of less than 10,000 ohm-centimeters) will have a direct effect on the resistance-to-ground of a grounding electrode (system). This is certainly desirable. However, where copper conductors (ground ring) and copper or copper-clad ground rods are installed in highly conductive soil and where these conductors or rods are in close proximity to a galvanized steel building foundation or tower support, this may lead to corrosion of the steel due to electrolytic action. For this reason, under these conditions, the use of galvanized steel rods, stainless-steel rods, and tinned copper conductors may be a better option.

The ground rod is not to be less than 8 feet in direct contact with the earth (2.44 m) and, typically 5/8" (15.87 mm) in diameter, although "listed" (UL 467) ground rods may be 1/2" (12.7 mm) in diameter note that NFPA 780 requires the ground rod(s) to the 10 feet (3 m) deep (where possible), so the top of the 8 foot long ground rod will be 2 feet below the earth surface.

A 3/4" trade size (metric designator 21) galvanized pipe may be used, with the same minimum length of 8 feet (2.44 m), in direct contact with the earth. However, the service life of this grounding electrode may be limited, possibly to no more than 10 years, depending on the soil resistivity.

6) Other Listed Electrodes are also permitted, such as a chemically charged ground rod (UL 467J). These are hollow core rods, and they are filled with various low-resistivity compounds, such as magnesium sulfate. The rod may have an outside diameter of 2-1/8" (53.975 mm), and they are normally 10 feet (3.048 meters) in length. However, they can be made longer, and I have used them in an "L" shape configuration. The compounds react with the normal moisture in the air, dissolve, and will slowly migrate into the soil through a series of holes in the shaft of the rod, thereby adding minerals to the soil and lowering the soil resistivity. The performance of this rod will increase over time, as the compounds within the rod migrate into the soil.

Introduction 23

These rods may have a service life of up to 50 years. Where the soil resistivity has not been tested, I would strongly recommend that these electrodes be considered.

7) Plate Electrodes. Plate electrodes, if of coated iron or galvanized steel, are required to be at least 1/4" (6.4 mm) in thickness. If of nonferrous material, they may be 0.06" (1.5 mm) in thickness. They must expose at least 2 square feet (0.186 m^2) to exterior soil. These plate electrodes, while not commonly used, are typically buried on end, as opposed to lying flat, in order to reduce the size of the excavation. 250.53(A)(5) requires this metal plate to be not less than 30 inches below the surface of the earth. As a comparison to the surface contact area of a ground rod, a one square foot metal plate has a surface contact area of 288 square inches and a 5/8 inch diameter ground rod that is 8 feet long has a surface contact area of 29.45 square inches. However, the surface contact area is negligible with respect to the resistance-to-ground of these grounding electrodes. The soil resistivity will determine the resistance-to-ground of these electrodes and the depth of the metallic electrode will be the deciding factor, as the soil will normally have a lower resistivity with increased depth. As a rule of thumb, doubling the depth of a ground rod will reduce its resistance-to-ground by approximately 40 percent.

250.52(B) addresses the materials and systems not permitted to be used as grounding electrodes.

1) A metal underground gas piping system. This is not necessarily due to an explosion hazard. It is because of the impressed current cathodic protection system that the gas supplier may be using to protect the gas pipe from galvanic corrosion. Connecting a grounding electrode conductor to the gas pipe could negate this protection due to the low level of ac current flowing through the grounding electrode conductor and into the gas pipe and then through the earth on its path back to the supply transformer.

Also, the gas pipe will typically have insulating joints. So, a long uninterrupted metal pipe may not be available and there will be a dielectric coupling in this pipe as it extends from the earth and before it enters the building.

2) Aluminum electrodes. Due to corrosion and oxidation, these electrodes would have a limited service life.
3) Structures and reinforcing steel that form the equipotential bonding for a swimming pool (680.26(A)(B)), the purpose of which is to reduce the

24   *Introduction*

voltage gradients (differences) in the pool area. This grid or structure is not a grounding electrode and it may prove to be hazardous to persons in the pool area if it were a part of the grounding electrode system, due to inadvertent voltage variations that are associated with the grounding electrode system.

Section 250.53(A)(2) states that a single rod, pipe, or plate is to be supplemented by an additional grounding electrode, and the supplemental electrode may be any of the types referenced in Sections 250.52(A)(2) through (A)(8). There is an exception here that waives this requirement if the single rod, pipe, or plate has a resistance-to-ground (earth) of 25 ohms or less. I have always maintained that this 25-ohm value is too high to provide proper protection for people, systems, and equipment, especially if there was a problem with lightning. If the service supply was affected by an indirect lightning strike, or more likely at an outside transformer supplying this distribution system, and the lightning current of, say, 20,000 amperes, which is quite low, was carried into this ground rod with a 25-ohm resistance-to-ground, the voltage-rise, above earth potential, would be 500,000 volts. While this would be an instantaneous event (lightning currents reach a peak in about 2–10 microseconds), this, certainly, would be hazardous to people and damaging to equipment. It should be noted that the 25-ohm resistance-to-ground of the ground rod would limit the level of fault current that may be carried into the earth. But, in order to effectively dissipate this current more effectively and reduce the voltage-rise above earth potential on systems and equipment, a lower resistance-to-ground electrode is important.

It is extremely important that where multiple grounding electrodes are installed, it is extremely important that they be bonded together to form one common grounding electrode system (250.50), (250.58), (250.66), (250.106), (800.100). This includes the grounding electrodes that are installed for the strike termination devices of a lightning protection system (250.60), (250.106). The bonding of all separate grounding electrodes will serve to limit potential differences between them and the systems and equipment that are connected to ground through these grounding electrodes. Also, the bonding of electrodes of different systems is addressed in (800.100(B)), (810.21(J)), (820.100(A)), (830.93(A)), (840.101(A)).

Another benefit associated with the bonding of the grounding electrodes of different systems is that the overall resistance-to-ground of all of the connected systems may be markedly reduced, as these systems will now be in parallel. This will provide added protection for all of these systems (250.50), (250.58).

Introduction 25

- **Grounding Electrode Conductor-**

The grounded conductor and the equipment grounding conductor are connected to the grounding electrode (system) by the grounding electrode conductor. Section 250.62 states that this conductor may be copper, aluminum, or copper-clad aluminum. And the conductor may be solid or stranded, insulated, covered, or bare.

Generally, this conductor must be installed in one continuous length. However, exothermically welded connections or compression (crimped) type connections are acceptable, as these connections are irreversible, which means that they cannot be taken apart with tools. Busbar connections are also acceptable (250.64(C)).

Where multiple service disconnects are installed as a group in separate enclosures (230.71), taps from each disconnect to the main grounding electrode conductor are permitted. This would be common where these separate enclosures are connected to an auxiliary gutter (Article 366), with the splices inside this gutter (250.64(D)(1),(2),(3)).

Enclosures for grounding electrode conductors should be nonmetallic or nonferrous. However, steel enclosures are acceptable for physical protection if this enclosure is bonded at both ends to the internal grounding electrode conductor. This may be accomplished through the use of bonding bushings on each end of the steel enclosure. The size of the bonding jumper will be the same size or larger than the internal grounding electrode conductor. The reason for this requirement is due to the magnetic properties of the steel enclosure. If significant current is flowing through the grounding electrode conductor, the magnetic field surrounding this conductor will induce a current in the steel enclosure, which is in a direction opposite to the current flow through the grounding electrode conductor. This condition will produce an "inductive choke" on the grounding electrode conductor, which will significantly increase the impedance of the earth connection and possibly destroy the grounding electrode conductor in the process 250.64(E), due to arcing within the steel enclosure.

When the steel enclosure is bonded to the internal conductor on both ends, the effects of this "inductive choke" are greatly reduced but not completely eliminated.

The steel enclosure (conduit) may be connected by means of grounding-type locknuts or bonding-type bushings with properly sized bonding jumpers (250.92(B)).

If nonferrous enclosures are used for the physical protection of the grounding electrode conductor, bonding the conductor on each end of the

26 *Introduction*

enclosure is not necessary, due to the nonmagnetic properties of this material (250.64(E)(1)).

The sizes of AC grounding electrode conductors are listed in Table 250.66 (for direct current systems, 250.166 applies). This size is based on the size of the ungrounded service-entrance conductors, or the size of the ungrounded feeder conductors that extend from the source of a separately derived system or the first disconnect or overcurrent device supplied from the source (250.30(A)).

A No. 6 AWG copper (13.3 mm^2) or No. 4 AWG aluminum or copper-clad aluminum (21.66 mm^2) conductor may be used as the sole connection to a ground rod, pipe, or plate electrode (250.66(A)). However, the use of an aluminum or copper-clad aluminum conductor for this purpose is limited due to the fact that aluminum or copper-clad aluminum conductors external to buildings or equipment enclosures are not to be terminated within 18 inches (45.72 cm) of the earth (250.64(A)), unless installed in outdoor enclosures that are listed and identified for the environment (250.64 (A)(2)).

For concrete-encased electrodes, the connection to the reinforcing steel within a concrete footing may be No. 4 AWG copper (21.66 mm^2) (250.66(B)). However, be sure that the connecting device is listed and identified for this connection, that is, copper to steel, or an exothermically welded connection may be used 250.70(A)).

Grounding electrode conductors in direct contact with the earth and buried for physical protection are not subject to the minimum cover requirements specified in 300.5 or 305.15 (250.64(B)(4)).

• Grounded, Functionally-

A system that has an electrical ground reference for operational purposes that is not solidly grounded. A PV system with DC circuits that exceed 30 volts and 8 amperes is required to have DC ground-fault protection. The ground-fault detector-interrupter is normally located within the inverter or charge controller (690.41(B)), (690.42). The grounded conductor (neutral) from the AC side of the inverter is referenced to ground at this location, or further downstream possibly at the location of the service equipment.

In the most common cases today where the PV system has an interactive inverter output, the equipment grounding system is connected to ground through the same grounding electrode system as the grounded AC distribution system at the service equipment.

Introduction 27

Interactive Mode-

The operating mode for power production equipment or microgrids that operate in parallel with and can deliver energy to an electric production and distribution network or other primary source (Article 100).

Inverter, Multimode-

Inverter equipment capable of operating in both interactive and island modes (705.50).

Inverter, Stand-alone-

Inverter equipment having the capabilities to operate only in island mode (Article 100).

Microgrid-

An electric power system capable of operating in island mode and capable of being interconnected to an electric power production and distribution network or other primary source while operating in interactive mode with the ability to disconnect from and reconnect to a primary source and operate in an island mode (Article 100).

Microgrid Control System-

A structural control system that manages microgrid operations, functionalities for utility interoperability, islanded operations, and transitions.

Microgrid Interconnect Device-

A device that enables a microgrid system to separate and reconnect to an interconnected primary power source (Article 100).

Panelboard, Enclosed-

Buses and connections with overcurrent protective devices in a cabinet or enclosure suitable for a panelboard application (Article 408-408.6, 408.30, 408.36, 408.37, 408.38, 408.39, 408.40, 408.41, 408.43, 408.50, 408.54, 408.55).

PV DC Circuit-

The PV dc circuit conductors between modules in a PV string circuit, and from PV string circuits or dc combiners, to dc combiners, electronic power converters, or a dc PV system disconnecting mean or dc-to-dc converter circuits.

28 *Introduction*

PV DC Circuit, String-

The PV source circuit conductors of one or more series-connected PV modules (Article 100).

PV (Photovoltaic) System-

The total components, circuits, and equipment up to and including the PV system disconnecting means.

Service Conductors-

These are the conductors which extend from the Service Point to the service disconnecting means (Parts II and III, Article 230).

Service Conductors, Overhead-

The overhead conductors between the service point and the first point of connection to the service-entrance conductors at the building or other structure.

Service Conductors, Underground-

The underground conductors between the service point and the first point of connection to the service-entrance conductors in a terminal box, meter, or other enclosure, inside or outside the building wall.

Service Entrance Conductors-

The service conductors between the terminals of the service equipment to the service drop, overhead service conductors, service lateral, or underground service conductors.

Bonding

• Bonded-

Connected to establish electrical continuity and conductivity. Equipment may be bonded through the use of a bonding conductor or jumper. This conducting path may be established through one or more of the conducting paths that are referenced in 250.118. In addition, this conducting path must not only be reliable, but it must also be able to safely carry any fault current likely to be imposed (250.90).

• Bonding Conductor (Bonding Jumper)-

A conductor that ensures the required electrical conductivity between metal parts required to be electrically connected. This conducting path must be

Introduction 29

designed to reduce the effects of voltage differences between the connected equipment and be able to safely withstand the maximum fault-current that may be imposed. The length of the bonding jumper should be limited in order to reduce the effects of voltage differences.

- **Bonding Jumper, Main-**

The connection between the grounded circuit conductor and the equipment grounding conductor, or the supply-side bonding jumper, or both, at the service (250.28). This conducting path must be capable of carrying the ground-fault current that may be flowing through it until the appropriate overcurrent device safely clears the fault on an electrical system that is solidly grounded. The minimum size of this bonding jumper is in accordance with Table 250.102(C)(1), and in some listed equipment, this conductor may be in the form of a screw (green) or copper strap. Where this conducting path is through a wire, the minimum size is determined in accordance with 250.122. However, just as in the case of the other bonding conductors referenced here, the maximum ground-fault current and its duration and the operating characteristics of the circuit overcurrent device and the short-time current carrying capacity of the Main Bonding Jumper must be known to assure the integrity of this bonding jumper.

For ungrounded systems, there will be an equipment grounding conductor, which is sized in accordance with 250.118 and installed with the supply conductors. Just as in the case with those systems that are solidly grounded, the available ground-fault current will be calculated and the equipment grounding conductor is sized accordingly. This conductor will be bonded to the supply side of the service disconnect if the supply system is ungrounded (250.25(B)). Where the utility supply system is grounded, the grounding of systems permitted to be connected on the supply side of the service disconnect are in accordance with 250.24 (250.25(A)). This includes a connection to the terminal or bus within the service disconnecting means where the grounded service conductor is connected, this is also the point of connection of the Main Bonding Jumper (250.24(B) (250.28) (250.102(C)(1)).

The grounding (bonding) connection of the grounding electrode conductor to the grounding electrode (system) is in accordance with 250.24(A)(1), 250.25(A), 250.25(B), 250.66, 250.166, 250.50, 250.52, and 250.53.

If the transformer supplying the service is located outside of the building, at least one additional grounding connection is required at the transformer or at another location outside the building (250.24(A)(2)).

30 *Introduction*

For portable, vehicle-mounted, and trailer-mounted generators, the provisions of 250.34 apply. The bonding of the frame of the portable, vehicle-mounted, and trailer-mounted generator to a grounding electrode is not required (250.52) where the generator only supplies equipment mounted on the generator and, or, cord-and-plug connected equipment through receptacles that are mounted on the generator. The metal parts of equipment and the equipment grounding conductor terminals of the receptacles are connected to the generator frame (OSHA 1926.404 f 3i).

If the generator is a separately derived system and is required to be grounded (250.26), a bonding connection from the grounded conductor of the generator to the generator frame is required (250.34(C)).

The grounding electrodes that are present at each building or structure that are described in 250.52(A)(1) through (A)(8) are required to be bonded together to form the grounding electrode system (250.50),(250.58). The bonding of these separate grounding electrodes will serve to limit voltage differences between these systems and also to reduce the resistance-to-ground of the grounding electrode system (250.50),(250.58),(250.60), (250.106),(690.47(A)(B)),(800.100(D)),(810.21(J)),(820.100(A)),(830.93(A) (B)), (840.101))

The exposed non-current carrying metal parts of PV module frames, electrical equipment, and conductor enclosures of PV systems are required to be connected to a properly sized equipment grounding conductor (250.134 or 250.136). Where the devices and systems are used for mounting PV modules and used for bonding module frames, these devices are required to be listed, labeled, and identified for this purpose (690.43(A)(B).

If the generator is a separately derived system and is required to be grounded (250.26), a bonding connection from the grounded conductor of the generator to the generator frame is required (250.34(C)).

• Bonding Jumper, Supply Side-

This conductor is installed on the supply side of a service or within a service equipment enclosure or for a separately derived system, and it ensures the electrical conductivity between metal parts required to be electrically connected (250.30(A)(2)).

The minimum size of this conductor is in accordance with 250.102 (C)(1).

• Bonding Jumper, System-

The connection between the grounded circuit conductor and the supply-side bonding jumper or the equipment grounding conductor, or both, at a

Introduction   31

separately derived system. This conducting path serves the same purpose as a main bonding jumper, in that it must be capable of safely carrying the ground-fault current that may be imposed on it until the appropriate overcurrent device operates to safely clear the ground-fault current on a system that is solidly grounded. And similar to a main bonding jumper, its minimum size is determined by 250.102(C)(1). However, its actual size will be determined by the total ground-fault current that may flow through it and the duration of this fault current in accordance with the operating characteristics of the overcurrent device (clearing time and fault let-through current) on the faulted circuit (250.30(A)(1)), (250.28(A)(B)(C)(D)), (250.30(C)).

2

PV Source and Output Circuit Calculations

The following example applies to a PV source and output circuit and the calculations that apply to this type of installation.

First, we examine the module type to be used and other related information.

Cell type – polycrystalline silicon

Cell size – 5 inch (125 mm)

Number of cells and connection type

– 72 in series-maximum system – 1,000 Vdc

Electrical Information:

Maximum power voltage – 32.6 volts

Open circuit voltage – 41.3 volts

Maximum power current – 5.26 amperes

Short-circuit current – 5.72 amperes

System configuration:

16 modules per string (in series)

10 modular strings

−6 °F – lowest expected ambient temperature (23.78 °C)

+105 °F – maximum expected ambient temperature (40.56 °C)

The maximum voltage is determined by multiplying the open circuit voltage (41.3 volts) of each module by the number of modules in series.

34 *PV Source and Output Circuit Calculations*

41.3 V
$\underline{X16}$ (in series)
660.8 V

The open circuit voltage is the voltage with no load on the system.

Table 690.7(A) identifies the correction factors for ambient temperatures below 25°C or 77°F. As the ambient temperature decreases, the electrical resistance decreases (negative temperature coefficient) and the open circuit voltage increases.

The lowest expected ambient temperature in our example is –6°F (23.78°C). According to Table 690.7(A), the correction factor for this temperature is 1.20.

660.8 V
$\underline{X1.20}$
792.96 V

Based on this voltage, a UL 2579 listed fuse with a maximum voltage of 1,000 Vdc will be required.

It should be noted that where the module nameplate information includes the manufacturer temperature coefficient %/C, this information must be used and not the correction factors listed in Table 690.7(A) to calculate the maximum PV system voltage. For example, where the manufacturer's temperature coefficient is stated as $-0.34\%/°C$, the maximum PV source circuit voltage in our example is calculated as follows:

Module open circuit voltage – 41.3 voltage 16 modules per string

41.30 V $\times$ 1 + [(–6 °C – 25 °C) $\times$ –0.34%/°C]) $\times$ 16 modules in series

41.30 V (1 + [(–31 °C $\times$ 0.34%/°C]) $\times$ 16 modules in series

41.30 V $\times$ (1 + 14.50%) $\times$ 16 modules

41.30 V $\times$ 1.1450 $\times$ 16 modules = 756.62 V

PV voltage = 756.62 volts

In these examples, the PV voltage has been calculated at 792.96 volts using Table 690.7(A) and 756.62 volts using the manufacturer's nameplate open circuit voltage values. So the PV voltage is 792.96 V.

The conductors that are associated with a PV system are subjected to a variety of conditions that include elevated ambient temperature, direct

PV Source and Output Circuit Calculations 35

sunlight, and wet locations. PV wire is more suitable to these environmental conditions than the use USE-2 wire.

PV wire has a 90°C wet rating and a 150°C dry rating and is suitable for connecting PV modules. And USE-2 wire is designed and intended for use as underground service entrance cable and it is listed at 90°C in wet and dry locations and rated for 600 volts.

PV wire has a minimum conductor size of 18 AWG and USE-2 has a minimum conductor size of 14 AWG.

PV wire may be rated at 1,000 and 2,000 volts for systems operating at over 600 volts.

NEC 690.7 limits the DC circuit voltage to 1,000 volts where these circuits are on or in buildings. However, for PV DC circuits in or on one family and two family dwellings, the maximum voltage is limited to 600 volts. Where not located on or in buildings, listed PV equipment rated at a maximum of no greater than 1,500 volts is not required to comply with Parts II and III of Article 490.

PV wire has a thicker insulation or an outside jacket to provide protection against physical damage.

The currents produced by a PV system are continuous. A continuous load is defined in Article 100 as a load where the maximum current is expected to continue for 3 hours or more. Branch-circuit conductors supplying continuous loads are sized in accordance with 210.19(A)(1), that is, the conductor ampacity is based on the noncontinuous load plus 125% of the continuous load.

The branch-circuit overcurrent device rating is based on the noncontinuous load plus 125% of the continuous load (210.20(A)).

These provisions also apply to feeder conductors and a feeder overcurrent device in accordance with 215.2(A)(1a) and 215.3.

It should be noted that the branch-circuit conductor size shall have an ampacity not less than the maximum load to be served after the application of any derating or adjustment factors referenced in 310.15 (210.19(A)(1)(2)). This provision also applies to feeder conductors in accordance with 215.2(A)(1)(2).

Due to the fact that grounded (neutral) conductors do not terminate (normally) on overcurrent devices, the rules for sizing conductors supplying continuous loads do not apply to grounded conductors, unless the grounded conductor terminates in an overcurrent device that opens all of the circuit conductors simultaneously (240.22(1)(2)).

36 *PV Source and Output Circuit Calculations*

Conductors associated with PV systems are normally subjected to high ambient temperatures and the provisions of 310.15(B) will apply.

Conductors that are subject to high ambient temperatures will have an increase in resistance (positive temperature coefficient). Tables 310.15(B)(1)(1) and 310.15(B)(1)(2) identify correction factors where the ambient temperatures are other than 30°C or 86°F and 40°C or 104°F. These correction factors are applied to the normal conductor ampacity to either increase or decrease the conductor ampacity.

An additional consideration applies to raceways or cables that are exposed to direct sunlight on or above rooftops where the distance above the roof to the bottom of the raceway or cable is less than 3/4 inch (19 mm). In this case, a temperature adder of 33 °C or 60 °F will be added to the outdoor temperature and this will determine the ambient temperature for the application of the listed correction factors in Table 310.15(B)(1)(1) or Table 310.15(B)(1)(2). This correction factor does not apply to the use of XHHW-2 insulated conductors (Table 310.15(B)(2), Exception).

To determine the ambient temperatures in various locations, consult the ASHRAE Handbook-Fundamentals (2017).

In addition to the ambient temperature correction factors of 310.15(B), there is another adjustment factor where more than three current-carrying conductors are installed in a raceway or cable or where single conductors or multiconductor cables are not in raceways and installed without maintaining spacing. Where the continuous length is longer than 24 inches (600 mm), the adjustment factors of Table 310.15(C)(1) will apply.

For example, where the temperature rating of the conductor is 75°C and the highest ambient temperature is determined to be 106°F and there are six current-carrying conductors installed in a rigid metal conduit, the correction and adjustment factor will be 0.82 from Table 310.15(B)(1)(1) and 0.80 from Table 310.15(C)(1). Combining both factors and multiplying the single factor by the conductor ampacity will determine the corrected ampacity (0.82 × 0.80 = 0.656).

This adjustment factor from Table 310.15(C)(1) does not apply to auxiliary gutters (366.23(A)) and metal wireways (376.22(B)) unless the number of current-carrying conductors exceeds 30 at any cross-section. For non-metallic wireways, the adjustment factors of 310.15(C)(1) apply to the current-carrying conductors up to the 20% fill expressed in 378.22.

Also, the ampacity adjustment factors of Table 310.15(C)(1) do not apply where the conductors are installed in a raceway having a length not exceeding 24 inches (600 mm) (310.15(C)(1)(b)).

PV Source and Output Circuit Calculations 37

Adjustment factors of Table 310.15(C)(1) do not apply to Type AC or Type MC cable if these cables have no outer covering, each cable has no more than three current-carrying conductors, the conductors are 12 AWG copper and not more than 20 current-carrying conductors are installed without maintaining spacing, and they are stacked or are supported on "bridle rings."

Neutral conductors that carry only the unbalanced current from conductors of the same circuit are not counted as current-carrying conductors when applying the provisions of 310.15(C)(1) (310.15(E)(1).

Information technology equipment, electric, electronic discharge lighting, adjustable-speed drive systems, and other equipment with switching power supplies are examples of nonlinear loads. These types of loads produce harmonic currents, primarily the third harmonic, on three-phase, four-wire wye-connected systems. The 180-cycle currents do not cancel and become additive in the neutral conductor. In some cases, the neutral current may exceed the currents in the phase conductors, which may require increasing the neutral conductor size and supplying the three-phase load through a properly sized "K" rated transformer.

For this reason, 310.15(E)(3) requires the neutral conductor to the counted as a current-carrying conductor and the 80% ampacity adjustment factor of Table 310.15(C)(1) applies.

310.15(F) – A grounding or bonding conductor is not considered to be a current-carrying conductor.

690.8(A)(1)(a)(1) – The maximum PV current is equivalent to the sum of the short-circuit current ratings of the PV modules that are connected in parallel multiplied by 125%.

In our example, the short-circuit current rating of each module is 5.72 amperes and we have 16 modules *in series* (strings) and we have 10 modular strings connected in parallel.

$$
\begin{array}{ll}
5.72 & \text{amperes} \\
\underline{\times 10} & \\
57.20 & \text{amperes} \\
\underline{\times 1.25} & \\
71.50\ \text{A} &
\end{array}
$$

The open circuit voltage has been calculated at 792.96 volts, or 793 volts. At 72 amperes and the use of 90°C insulation, a 6 AWG copper conductor may be used if the equipment terminations are also identified at 90°C (75 amperes at 90°C) – 110.14(C)(1)(a)(3) – Table 310.16.

38   *PV Source and Output Circuit Calculations*

In our example, the equipment terminations are identified at 90°C. If the equipment terminations are identified at 75°C and the conductor size is 14 AWG through 1 AWG (which it is in our example), the conductor ampacity is based on the 75°C ampacity of the conductor is accordance with 110.14(C)(1)(a)(3) and the conductor size will be 4 AWG copper (85 amperes) (Table 310.16).

The source circuit current is calculated by multiplying the short-circuit current by 1.25.

$$\begin{array}{r} 5.72 \text{ amperes} \\ \times 1.25 \\ \hline 7.15 \text{ amperes} \end{array}$$

Before the application of adjustment or correction factors, the conductor ampacity for this installation is based on the following:

$$\begin{array}{r} 7.15 \text{ amperes} \\ \times 1.25 \\ \hline 8.94 \text{ amperes} \end{array}$$

Referring to Table 310.16, it appears that a 90 °C rated conductor size would be 18 AWG.

However, 240.4(D)(1) limits the overcurrent protection for this conductor size to no more than 7 amperes. The maximum circuit current has been determined at 8.94 amperes and the next standard overcurrent device rating is 10 amperes (240.4(B)(240.6(A)). A 14 AWG copper conductor may be used in accordance with 240.4(D)(3).

The ambient temperature correction factor, based on the 105°F maximum ambient temperature and the values expressed in Table 310.15(B)(1)(1) is 0.87.

$$\frac{7.15 \text{ amperes}}{0.87} = 8.22 \text{ amperes}$$

This ampere value (8.22 A) is less than the calculated 8.94 A (maximum circuit amperes); so the larger value must be used (690.8(B)(1), 690.8(B)(2)).

Based on the 14 AWG copper PV wire and the maximum circuit amperes of 8.94 amperes, the next larger fuse size of 10 amperes is to be used (240.4(B)) (690.9(A)). This is a 10 A PV fuse rated at 1,000 V.

The next step is to determine the maximum output current. Each string has a maximum current of 7.15 amperes and there are 10 parallel strings. So, the maximum output current is 71.50 amperes × 1.25 = 89.375, or 89.40 A.

PV Source and Output Circuit Calculations 39

Based on the use of 90°C rated insulation, a 4 AWG copper conductor (95 A) may be used, as we have determined that the equipment terminations are rated at 90°C.

Next, the conductor size is based on the ambient temperature correction factor (0.87), divided into 71.50.

Divided into 71.50 amperes

$$\frac{71.50 \text{ amperes}}{0.87} = 82.18 \text{ amperes}$$

The conductor size is based on the larger of 89.40 A or 82.18 A. In either case, a 4 AWG copper conductor with a 90°C ampacity of 95 A can be used.

The overcurrent device for this circuit is based on 71.50 A × 1.25 = 89.40 A. The next higher overcurrent device rating from Table 240.6(A) is 90 amperes (240.4(B)).

The ampere and voltage rating for this fuse will be a 90-A PV fuse at a maximum voltage of 1,000 V.

Of course, in this PV system example, there are other considerations, such as voltage-drop and system grounding and bonding.

Voltage-drop may be calculated by the use of the following formulas:

$$\text{voltage} - \text{drop} = \frac{2K \times L \times I}{\text{cm}} - \text{single} - \text{phase or dc}$$

$$\text{voltage} - \text{drop} = \frac{1.732K \times L \times I}{\text{cm}} - \text{three} - \text{phase}$$

K = 12.90 ohms-copper

K = 21.20 ohms-aluminum

L = one-way length in feet of conductor

I = amperes of load

cm = circular mil area of conductor (Table 8, Chapter 9).

Of course, the voltage-drop may have an effect on the circuit conductor size (210.19(A), Informational Note (215.2(A)(2), Informational Note No. 2).

By far, most PV systems are functionally grounded. And, as this term is defined in Article 100, this system has an electrical ground reference for operational purposes, which is not solidly grounded. In most installations, a ground-fault detector-interrupter is installed within the inverter. This device is not provided for people protection, as in the case of ground-fault circuit interrupters, but to protect against fire. The grounding electrode (system) on

40 PV Source and Output Circuit Calculations

the alternating current side of the inverter connects the ac grounded conductor and the equipment grounding conductor to ground and provides the ground reference for the ground-fault protection. Section 690.41(B) requires DC ground-fault detector-interrupter for PV systems that exceed 30 volts or 8 amperes. PV systems that are solidly grounded and the PV source circuits have not more than two modules in parallel and are not on or in buildings are permitted without ground-fault protection. As will be the case in virtually all electrical installations, exposed non-current-carrying metal parts of PV module frames, electrical equipment and conductor enclosures are required to be connected to an equipment grounding conductor (250.134−250.136). The ground-fault current should be calculated, as well as the fault current in accordance with the information in Chapter 1 of this book. The equipment grounding conductor may be sized from 250.122, and normally, where the ungrounded conductors are increased in size to compensate for circuit conditions, the equipment grounding conductor is proportionately increased in size as well (250.122(B)). However, this does not apply where the circuit conductors are increased in size to compensate for higher ambient temperatures (310.15(B)(1)(2)) and/or where the circuit conductors are increased in size because there are more than three current-carrying conductors installed in a raceway or cable (310.15(C)). Also, where the circuit conductors are increased in size to compensate for *voltage-drop*, the equipment grounding conductor must also be proportionally increased in size. However, this does not apply to PV systems, including the PV modules, supporting metal frames, and equipment secured to grounded metal supports (690.45). A low resistance-to-ground grounding electrode system is essential, as well as short grounding electrode conductor connections that will serve to limit the voltage rise above earth potential on the entire system.

PV System-Grounding and Bonding

Where the DC circuits of PV systems exceed 30 volts or 8 amperes, they must be provided with DC ground-fault detector-interrupter protection in order to reduce fire hazards.

As we have previously stated, the ground-fault protection is normally included within the inverter, charge controller, or DC-to-DC converter. If this equipment does not have this protection, it is common that a statement is included in the equipment manual that indicates that a ground-fault detection device is not provided.

PV Source and Output Circuit Calculations 41

The ground-fault protection device or system is required to detect ground-faults in the PV system DC circuit conductors, including any functionally grounded conductors and be listed for this purpose (690.41(B)(1)).

Faulted circuits shall be controlled by one of the following methods, (690.41(B)(2).

1) The current-carrying conductors of the faulted circuit shall be automatically disconnected.
2) The device providing ground-fault protection fed by the faulted circuit shall automatically cease to supply power to the output circuits and interrupt the faulted PV system DC circuits from the ground reference.

The ground-fault protection equipment shall indicate a ground-fault at a readily accessible location (690.41(B)(3)).

Section 690.43 requires that exposed non-current-carrying metal parts of PV module frames be connected to an equipment grounding conductor in accordance with 250.134 or 250.136. This provision includes any of the equipment grounding conductors that are identified in 250.118, or by the connection to an equipment grounding conductor of the wire type that is contained within the same cable, same raceway, or otherwise run with the circuit conductors (690.43(C)).

Section 250.136 recognizes that where equipment is secured to and in electrical contact with a metal rack or structure for its support, this equipment is considered to be connected to an equipment grounding conductors if the metal rack or structure is connected to an equipment grounding conductor that is referenced in 250.118 or by means of connection to an equipment grounding conductor of the wire type that is contained within the same raceway, cable, or otherwise run with the circuit conductors.

The equipment grounding conductors for the PV array and support structure shall be contained within the same raceway or cable or otherwise run with the PV system conductors where those circuit conductors leave the vicinity of the PV array (690.43(C)). This is an example of 90.3, where 690.43(C) modifies 250.134, Exception No. 2, which permits the equipment grounding conductors for DC circuits to be run separately from the circuit conductors.

Equipment grounding conductors for PV system circuits are sized in accordance with 250.122. If no overcurrent protection is used, an assumed overcurrent device rating in accordance with 690.9(A)(1),690.9(B) shall determine the size of the equipment grounding conductor when applying Table 250.122.

42 *PV Source and Output Circuit Calculations*

Increasing the size of the equipment grounding conductor to compensate for voltage-drop is not required (690.45). This is another an example of 90.3, as 690.45 modifies 250.122(B).

If equipment is secured to grounded metal supports and if the metal support is connected to a properly sized equipment grounding conductor in accordance with 250.134, this equipment is considered to be connected to an equipment grounding conductor. If the metal support is painted or coated with enamel, 250.12 requires these coatings to be removed in order to maintain proper electrical contact (250.136).

Devices that are listed, labeled, and identified for bonding and grounding are permitted to bond PV equipment to grounded metal supports (690.43(B)).

The metal support structures for PV modules must have identified bonding jumpers installed between separate metallic sections or the metal support structure must be identified for equipment bonding and connected to the equipment grounding conductor (690.43(B)).

690.47(A) requires a building or structure supporting a PV system to have a grounding electrode system in accordance with Part III of Article 250.

For functionally grounded PV systems, there will be an equipment grounding conductor run with the AC output circuits from the inverter and this conductor is connected to the grounding electrode system. This ground connection is provided for the ground-fault protection as well as for the equipment grounding of the PV array.

If the metal frame of the building or structure is connected to ground in accordance with 250.52(A)(2), that is, where there is at least one metal in-ground support in direct contact with the earth vertically for 10 feet (3 m) or more, a roof-mounted PV array may use this metal frame as the required grounding electrode connection, (250.68(C)(2)).

Surge Protection

A Surge Arrestor may be installed ahead of a supply transformer so as to discharge the surge energy associated with lightning induced influences from utility transmission lines.

These transmission lines act as a large antenna in the presence of electrically-charged thunderclouds. Electrical charges are predominately direct current, although there is an alternating current component as well. These charges accumulate when warmer air from the earth surface rises and meets cooler air at higher elevations. This mixing of warmer and cooler air creates turbulence, which causes the rapid contact and separation of materials in the air (dust, dry ice, snow, etc.). The contact and separation of these materials causes static electricity to be produced within the cloud and pockets of negative and positive charges begin to form. As the cloud mass moves through the air, opposite charges are induced on the earth. And these earthbound charges follow the cloud.

A passive lightning prevention system recognizes that the function of the air terminals of a typical lightning protection system are not to attract lightning from the cloud, but, to discharge the opposite polarity earthbound charges at the points of the strike termination devices, thereby removing the difference of potential between a building or structure and the opposite charges in the cloud mass.

An <u>active lightning prevention system</u> recognizes that the sharp points of the strike termination devices form a corona cloud above them that makes these prevention devices less attractive to the lightning strike, thus preventing the strike from damaging the building or structure.

An active lightning attraction system is designed to attract the lightning strike. This concept is thought to intercept the lightning down stroke (stepped leader) through the ionized air above the strike termination devices and provide a more attractive path for the lightning energy to flow to ground (electrode).

Lightning can occur within the cloud, or from cloud to cloud, or from the cloud to the earth. As the positive and negative charges build within the cloud, the magnetic field associated with this charged mass begins to gain strength and the air surrounding the cloud begins to ionize. The ionized air begins to lose its dielectric properties and the charges are free to flow toward the earth through a series of strokes, each about 150 feet (45.72 meters) in length. This is known as a stepped leader and it creates a path that becomes closer to objects on the earth, especially at higher elevations or at isolated buildings or structures. It is at this time that the opposite earthbound charges may rise to meet the downward 'stepped leader', and a return stroke occurs. This is the bright streamer that we see and it flows from the earth to the cloud through the conducting path that was formed by the 'stepped leader'. This process may develop several times before the charges in the cloud and their opposite earthbound charges are neutralized.

Many years ago, I found information regarding thunderstorm activity in the U.S. and Canada. This was an 'isokeraunic map', and areas were categorized by thunderstorm day activity. The term 'isokeraunic' is derived from two Greek words, 'iso' meaning 'same', and 'keraunous' meaning 'thunderbolt.' A 'thunderstorm day' is identified as one where thunder is

44 PV Source and Output Circuit Calculations

heard at least once. Of course, topography plays an important role in this determination. The corridor between Tampa and Orlando, Florida is known as 'Lightning Alley', with as many as 90 'thunderstorm days' per year. Other areas, such as Southern California, have very few 'thunderstorm days'. This is a good parameter in determining the need for lightning or surge protection.

It is easy to see that exposed transmission and other power lines are a constant threat, due to the âĂŸantenna effectâĂŹ that they provide during thunderstorm activity. And the concern is not only due to a direct lightning strike, but also, of induced charges, that may affect these exposed conductors. A high voltage transient, and resultant surge current, could easily flow through these exposed lines. And the âĂŸvoltage waveâĂŹ may cause surge current to flow in both directions through these conducting paths. Add to this the fact that where a power line terminates, the voltage wave can double in magnitude. Any electrical service near the termination of this power line would need additional protection. This may be a lightning protection system for the building or structure (NFPA 780), with additional protection in the form of a âĂŸSurge ArresterâĂŹ, and one, or more, Surge Protective Devices downstream.

The System Grounding requirements for PV systems involve the following Grounding Arrangements (690.41)

1) Two-wire PV Arrays with one functionally grounded conductor.

For example, consider an arrangement of PV modules that generate dc power at the appropriate voltage and current. The dc circuit conductors between the modules and from the modules to dc combiners, electronic power converters, or a dc PV system disconnecting means, are the Photovoltaic Power Source Circuit, and it extends from the power source to a common connection point(s) of the dc system. When there are two or more PV Source Circuits, a Direct Current Combiner is used to combine these circuits and provide for one dc circuit output. The Photovoltaic Output Circuit extends from the PV source circuit to the Inverter or to the dc utilization equipment. The Inverter Input Circuit includes the conductors that are connected to the dc input of the inverter. The inverter is the equipment that converts the dc input to an ac output. The Inverter Output Circuit includes the conductors that are connected to the ac output of an Inverter and to the ac connected load.

A Functionally Grounded PV system is one where there is no solidly grounded conductor on the dc side of the Inverter. And the Equipment

PV Source and Output Circuit Calculations 45

Grounding Conductor from the Inverter ac output circuit(s) provides the ground connection for the Ground-Fault Detector-Interrupter and equipment grounding of the PV arrays. The EGC is connected to the grounding electrode (system) on the ac side of the Inverter.

2) Bipolar PV arrays, with a functionally ground reference (center tap). A Bipolar PV array has 2 dc outputs, each having an opposite polarity to a common reference point or center tap, which is where the ground (earth) connection is made (690.7(C)).
3) Circuits not isolated from the grounded inverter output circuit.
4) Ungrounded circuits
5) Solidly grounded circuits as permitted in 690.41(B).
6) Circuits that use other methods that accomplish equivalent system protection in accordance with 250.4(A), with equipment listed and identified for the use. Section 250.4(A) covers electrical systems that are grounded (solidly), as well as the bonding of electrical equipment, and a means of providing an effective ground-fault current path, which for a solidly grounded system, will facilitate the operation of the circuit overcurrent device during a ground-fault, or initiate an alarm on a high-impedance grounded system or an ungrounded system (250.36(2)),(250.21(B)).

Section 690.41(B) requires dc Ground-Fault Detector-Interrupter Protection for PV system dc circuits that exceed 30 volts or 8 amperes, with the exception of PV arrays with not more the 2 PV Source Circuits, where all PV system dc circuits are not on or in buildings and the system is solidly grounded. The Ground-Fault Protection must detect ground-faults in the PV array dc conductors and equipment, and the faulted circuits must be automatically disconnected, or the inverter must automatically stop the current flow to the output circuits, as well as isolate the PV system dc circuits from the ground reference in a Functionally Grounded System. The Ground-Fault Detector-Interrupter is not provided as a means of personnel protection. It is meant to provide protection against fire hazards.

The connection to ground for any current-carrying conductor is made by the GFDI for functionally grounded systems, or for solidly-grounded PV systems, the dc circuit grounding connection is made at any single point of the PV Output Circuit, that is, from the PV Source Circuit to the Inverter.

Several years ago, PV systems were solidly grounded, that is the dc Source Circuit negative conductor was connected to a grounding electrode (system).

46 *PV Source and Output Circuit Calculations*

The recommendation was to make this connection to ground at a point that was close to the dc sources (modules) to afford better protection from lightning. This is true for any supply system, dc or ac. Also, keep in mind that the PV source is not a constant voltage source due to variations in the ambient temperature. As the temperature decreases, the supply voltage increases (Table 690.7(A)).

The grounding electrode (system) tends to stabilize the system voltage to ground. This is not to say that the earth connection would limit this voltage-rise as the ambient temperature decreases.

The grounding electrode (system) would also be used as a means of grounding the exposed noncurrent carrying metal parts of PV module (support) frames, as well as the metal frames of electrical equipment and conductor enclosures.

The ac supply system from the inverter would also be connected to a grounding electrode (system). This would include the grounded conductor (neutral) and the equipment grounding system.

Of course, both systems, dc and ac, could be connected to the same grounding electrode (system), which would include the grounding electrodes referenced in 250.52(A) and 250.166. If the dc and ac systems are connected to separate grounding electrodes, they must be bonded to limit potential differences between the dc and ac systems (250.50), (250.58). Also, if the grounding electrode is a single ground rod, pipe, or metal plate, with a resistance-to- ground of over 25 ohms, this grounding electrode must be supplemented by an additional electrode, which may be an additional rod, pipe, or plate (250.53(A)(2)).

However, in more recent times, a Functionally Grounded System is the preferred method of system grounding. In this case, the dc system is not solidly grounded and the Equipment Grounding Conductor for the Inverter ac output, which is connected to the grounding electrode (system), either at the Inverter or further downstream, provides the necessary ground reference for the dc Ground-Fault Protection and the equipment grounding of the PV supply system. This includes the metal support structures of the PV arrays and any metallic enclosures that are a part of the dc supply. The Ground-Fault Detector-Interrupter will normally be part of the Inverter.

In areas of significant thunderstorm activity, it may be beneficial to connect the PV array(s) support structures to an auxiliary grounding electrode (system) as a means of lightning protection (250.52),(250.54). If used, it would not be necessary to bond the auxiliary grounding electrode, as the bonding is already afforded through the equipment grounding system. Or,

PV Source and Output Circuit Calculations 47

if the building or structure supports a PV array(s), 690.47(A) requires that the arrays be connected to a grounding electrode system that is installed in accordance with Part III of Article 250 (250.52(A)). For example, if the building or structure has a grounding electrode system consisting of metal in-ground support structures (250.52(A)(2)), the PV array support system will be bonded to this grounding electrode. In this way, the array equipment grounding conductors will have a connection to the grounding electrode system through the metal structural frame of the building (250.68(C)(2)).

EQUIPMENT GROUNDING

Just as in other electrical systems, the exposed noncurrent carrying metal parts of the PV equipment, including the PV module supporting frames, metal enclosures, and metallic raceways and cable assemblies are required to be grounded. Equipment grounding conductors must be installed within the same raceway or cable assembly or otherwise run with the circuit conductors (300.3(B)).

690.43(C) permits the equipment grounding conductor to be run separately from the PV system conductors within the PV array where the PV circuit conductors leave the vicinity of the PV array, the equipment grounding conductors shall comply with 250.134.

The size of the equipment grounding conductors for PV Source and PV Output Circuits are based on 250.122. This means that the EGC is sized in accordance with the size of the circuit overcurrent device.

However, the circuits may originate from the PV modules where there are no overcurrent devices protecting these circuits (690.9(A)(1)). In this case, an assumed overcurrent device, in accordance with 690.9(B), (690.9(C)), will determine the size of the equipment grounding conductor(s).

Section 690.9(B) requires that the overcurrent devices that are used in PV systems must be listed for use in PV systems. Electronic devices that are listed to prevent backfeed current in PV system dc circuits are permitted to prevent overcurrent of conductors on the PV array side of the device. This means that these devices must be listed and identified for dc systems, which excludes overcurrent devices that are listed only for ac systems. Alternating current passes through zero two times during each cycle. Direct current is constant, and, especially, during short-circuit or ground-fault conditions, these faults are more difficult to interrupt than in alternating current circuits (690.9(B).

48   *PV Source and Output Circuit Calculations*

For PV systems, the overcurrent devices must be rated at not less than 125% of the maximum currents calculated in accordance with 690.8(A)(1)(a)(1)),(690.9(B)(1)).

Or, if an assembly contains overcurrent devices that are listed for continuous operation at 100% of their rating, these overcurrent devices may be loaded to 100% of their rating (690.9(B)(2)).

690.8(A)(1)(a)(1)), states that the maximum calculated current is the sum of the parallel connected PV module rated short-circuit currents multiplied by 125%.

For Example:

10 series connected DC modules with a short-circuit current rating of 8.9 amperes, each.

8.9 amperes $\times$ 1.25 (125%)=11.125 amperes.

The maximum PV Source Circuit Current is 11.125 amperes (690.8(A)(1)) (a)(1)).

The overcurrent device rating will be 1.25 (125%) times 11.125 amperes, or 13.90 amperes to satisfy 690.9(B)(1)).

When the 125% value of 690.8(A)(1) is applied to calculate the maximum PV Source Circuit current, and this is combined with the overcurrent device rating of 125% from 690.9(B)(1), the result is 156% (1.25×1.25=1.5625). 8.9 amperes $\times$ 1.5625 =13.90 amperes

So, if the calculated short-circuit current from 690.8(A) and 690.9(B) is 13.90 amperes, the equipment grounding conductor may be 14 copper $(2.08mm^2)$ from Table 250.122. This applies whether, or not, the source circuits are provided with overcurrent protection.

The equipment grounding conductors for the PV Source Circuits and PV Output Circuits (the circuit(s) from the PV combiner to the Inverter), do not have to be increased in size to compensate for voltage-drop (690.45). These circuits are dc and are not subject to inductive reactance. The minimum size is No.14 (2.08mm2)(250.122(A)),(Table 250.122)).

For PV systems with a generating capacity of 100 kW or more, the PV Source Circuits may be determined through calculations made by a licensed professional electrical engineer. An industry standard that may be used to determine the maximum current of a PV system is 'Sand 2004-3535',

PV Source and Output Circuit Calculations 49

from Sandia National Laboratories. The use of this calculation method will typically result in a lower current value than the calculation method of 690.8(A)(1)(a)(1). However, the calculation from this standard must be not less than 70% of the value from 690.8(A)(1)(a)(1), 690.8(A)(1)(a)(2).

Any exposed noncurrent carrying metal parts of PV module frames or panels and electric equipment enclosures must be grounded, regardless of voltage (690.43), (250.134), (250.136)). Metallic mounting structures (not building steel) may be identified as equipment grounding conductors and identified bonding jumpers or devices may be used to provide an acceptable bonding means between separate metallic sections. Equipment mounted on these structures (modules) may be bonded with listed devices identified for this purpose. The structure must be bonded to the equipment grounding system (690.43(A),(B)).

For PV modules, the provisions of 250.120(C) and 250.134(B), Exception No. 2 may apply. In this case, the equipment grounding conductor may not be routed with the circuit conductors. And where the equipment grounding conductors are smaller than No. 6 AWG (13.3 mm2), physical protection, in the form of a raceway or cable armor is required, unless the conductor is installed in such a way that physical protection is not necessary.

This information is another example of the Code Arrangement of Section 90.3, where Chapters 1-4 apply, except as amended by Chapters 5-6 and 7. Section 300.3(B) requires that conductors of the same circuit, including equipment grounding conductors, must be contained in the same raceway, cable, auxiliary gutter, cable tray, cable bus assembly, cable, or cord in order to limit overall circuit impedance. Section 300.5(I) identifies the same requirement for underground systems.

But, once again, the equipment grounding conductors from the PV modules are not subject to ac impedance, only dc resistance. So, installing these conductors as single conductors, with physical protection as necessary, is not a problem where these conductors are properly sized.

GROUNDING ELECTRODE SYSTEM

The grounding of AC modules, or the grounding of the ac system from an Inverter to a grounding electrode (system) will be in accordance with 250.50 through 250.60. Any ferrous metallic enclosure for a grounding electrode conductor must be bonded (on both ends) to the internal grounding electrode conductor to comply with 250.64(E)(1).

50   *PV Source and Output Circuit Calculations*

Solidly Grounded Systems-

For solidly grounded systems, it is common to have photovoltaic systems with both dc and ac grounding requirements. The dc grounding system must be bonded to the ac grounding system in order to limit voltage differences between these systems. This is a similar requirement to 250.50 and 250.58, and for the same reason. If two grounding electrode conductors are installed, one for the dc system and one for the ac system, a bonding conductor will be installed between these systems. The bonding conductor size is based on the larger size of the ac grounding electrode conductor or the dc grounding electrode conductor (250.66), (250.166).

Or, a dc grounding electrode conductor, sized in accordance with 250.166, may be installed from the dc grounding electrode connection point to the ac grounding electrode. It is also common on PV systems to have an isolation transformer installed in order to separate the dc grounded circuit conductor from the ac grounded circuit conductor. Once again, there must be a bonding connection between these systems. Or, these systems may be connected to a common grounding electrode in order to establish the same zero-volts potential reference to the earth.

Any auxiliary grounding electrode (Section 250.54) installed at ground level for roof-mounted or pole-mounted arrays must be connected to the array frame or supporting structure. It may be possible to bond the frame of a roof-mounted array to the metal frame of a building, if the metal building frame is part of the grounding electrode system (250.52(A)(2))(690.47(B)) (250.68(C)(2)).

It should be noted that the continuity of the equipment grounding system must be maintained, even if equipment is removed for repair or replacement. The PV source and PV output circuits remain energized as long as the PV modules are exposed to light. For example, if the inverter is removed for service, a bonding conductor must be installed to maintain the connection to the grounding system.

In addition, if the removal of equipment causes the Main Bonding Jumper in the Inverter to be disconnected, a bonding jumper must be installed to assure the grounding connection to the system grounded conductor. There may be a significant voltage-rise on the grounded conductor if the connection to the grounding electrode is interrupted.

A 'Functionally Grounded PV System' has an electrical reference to ground through an element (fuse, circuit breaker, or electronic device) that is a part of a listed ground-fault protection system (GFDI) and that is not solidly grounded. These systems will be at earth potential under normal conditions, but will be at an elevated voltage above earth potential during fault conditions.

BACK-UP GENERATORS

We have stressed throughout this book that the bonding of the grounding electrodes of different systems is of critical concern due to the significant voltage differences between these systems, which may, at times, be significant and in some cases, even lethal. A case in point was a newspaper article from many years ago that involved a young man living in suburbia who was using the telephone (landline) during a thunderstorm. Later, one of his parents found him deceased in his room. It was determined that the communications system was indeed bonded to a ground rod. However, this ground rod was not bonded to the grounding electrode for the service. Whether the cause of the voltage difference between these systems was a voltage-rise on the telephone system or some electrical device supplied from the service equipment was not known. However, it is obvious that this person had made contact with both of these supply systems at exactly the time when one system was affected by a significant voltage-rise, either direct or induced, while the other system was held at virtual earth potential. Incidents, such as the unfortunate death of this young man, have occurred many times. And, sometimes the cause of the accident goes undetected. I must emphasize that even with the proper bonding of the grounding electrodes of different systems (250.50, 250.58, 250.60, 250.94, 250.106, 800.100, 800.106, 810.21, 820.100, 830.93, and 840.101), there will be a voltage difference between connected systems, which may be significant due to the inductive voltage-drop of the bonding conductor connecting these systems. We have stated previously in Chapter 1 that lightning currents have frequencies that range from 3 kHz to 10 mHz. Due to the increased frequency and its effect on the bonding conductor impedance, the voltage difference between these connected systems may be significant. So, the length of the bonding conductors must be limited and unnecessary loops and bends in these conductors must be avoided.

This leads us to the installation of back-up generators for solar PV systems, emergency systems, optional standby systems, and, in fact, any back-up system. It is critical to locate the back-up supply system as close as possible to the normal supply system, and each system is connected to ground with conductors that are as short and straight as possible (250.4(A)(1), Informational Note No. 1). A "0" volts reference point is established for each system. The ground potential difference between the separate systems must

52　PV Source and Output Circuit Calculations

be limited and the overall connected electrical load needs to see the same ground (earth) potential in order to function properly, regardless of whether the load is supplied from the electrical service or the back-up generator. And with the back-up system installed close to the service, it is likely that the same grounding electrode (system) for both systems may be used.

Article 242 – Overvoltage Protection

Generally, surge-arresters are a part of the utility supply system. In areas of significant thunderstorm activity, or where topography may be a factor in deciding on the need for additional protection in the form of surge-arresters on supply systems of over 1,000 volts (NEC Article 242), we have to recognize that it is not uncommon to see voltages of several thousand volts on 120-volt premises wiring systems. We recognize that exposed utility lines act as an antenna for induced voltages and surge currents associated with direct, or even nearby (indirect) lightning strikes. In addition, utility switching operations often lead to significant overvoltages. The standard for surge-arresters is ANSI/IEEE C62.11. A review of this standard as well as NEC Article 242 is highly recommended.

As the transient surge passes through the arrester, a wave will flow in both directions, that is, into the distribution system and into the building supply system (transformer). Therefore, the surge-arrester must be installed in close proximity to the transformer and referenced to ground through conductors that are as short and straight as possible (242.48). This will significantly reduce the transient rise. This transient wave will increase in magnitude as the distance from the arrester also increases.

The surge arrester rating is not less than 125 percent of the maximum continuous operating voltage at the point of application. If the system is solidly grounded, this will be the phase-to-ground voltage. For impedance or ungrounded systems, this is the phase-to-phase voltage (242.42).

Some important aspects of providing protection through the use of surge-protective devices should be noted here. Keep in mind that these devices, while carefully designed and listed (UL 1449), can, and do fail, usually in a short-circuit mode. Providing overcurrent protection for these devices, as some manufacturers suggest, may be a good idea, especially where the equipment is frequently monitored and the source of a failed component may be promptly corrected.

In addition, when a voltage variation causes the operation of the SPD, a large surge current will be directed into the grounding conductor to which the

PV Source and Output Circuit Calculations 53

SPD is connected. There will be a momentary voltage-rise on this grounding conductor, which may be significant. Any other equipment that is connected to this grounding conductor will also be affected by this condition. The best method of limiting this momentary voltage-rise is to provide a grounding electrode system that has a low resistance-to-ground.

The voltage transient or variation that the SPD is designed to "clip" should be set at a value that is above the peak line-voltage, or 1.414 of the effective (rms) value of the voltage. For example, for a nominal voltage of 120 volts, the SPD should have a protective level that is *above* 170 volts (120 × 1.414 = 169.68 volts).

The upper limit of the effective voltage level (rms) that the SPD can safely withstand will be identified as the maximum continuous operating voltage (MCOV).

An addition to the 2014 NEC requires that a listed surge-protective device be *in* or *on* all emergency system switchgear, switchboards and panelboards (Section 700.8).

The SPD must be a listed device (UL 1449). The listing may permit the SPD to be installed on ungrounded, impedance grounded, or corner-grounded delta systems if the SPD is specifically listed for these types of systems (242.6),(242.12(2)).

An SPD must provide indication that it is functioning properly (242.9). The SPD must be identified with a short-circuit current rating, which must be equal to or in excess of the calculated available fault current (242.8).

Type 1 SPDs may be installed on the supply side of the service disconnect overcurrent device, the load side of the first overcurrent device at a building or structure, or on the load side of the first overcurrent device supplied by a separately derived system (242.13(A)(1)(2)).

The Type 1 SPD is connected to the grounded service conductor, the grounding electrode conductor, the grounding electrode for the service, or the equipment grounding terminal in the service equipment (242.13(B)).

Due to the high-frequency currents associated with lightning discharges, the conductors that connect the SPD to ground must not be any longer than necessary to make the connection and shall avoid unnecessary loops or bends. Lightning currents have frequencies that range from 3 kHz to 10 mHz. These high frequencies and the increased impedance associated with these high frequencies will cause longer conductors to be ineffective, as the clamping voltage of the SPD will be driven higher, thereby compromising the protection afforded by the SPD. This also applies to grounding electrode conductors which must have a limited length in order to hold systems and

54   PV Source and Output Circuit Calculations

equipment at close to earth potential ("0" volts), (250.4(A) Informational Note. 1).

Type 2 SPDs may be connected anywhere on the load side of the service disconnect overcurrent device (242.14(A)). Where a building or structure is supplied by a feeder, the Type 2 SPD may be connected anywhere on the load side of the first overcurrent device at the building or structure (242.14(B)). The Type 2 SPD may be connected on the load side of the first overcurrent device supplied by a separately derived system (242.14(C)).

A Type 2 SPD is permitted on the supply side of a service disconnect where the SPD is installed as part of listed equipment, which includes a disconnecting means and properly sized overcurrent protection (242.14)(A),(230.82(8)).

Type 3 SPDs may be connected on the load side of a branch-circuit overcurrent device up to the equipment served. The manufacturer may provide an instruction that states that the Type 3 SPD connection must be at least 30 feet (10 m) of conductor distance from the service or separately derived system disconnect (242.16).

Type 4 SPDs are installed by the equipment manufacturer (242.18).

Line and grounding conductors must not be smaller than 14 AWG copper or 12 AWG aluminum (2.08−3.31 mm^2) (242.28).

A Type 1 or Type 2 SPD must be provided for all services supplying dwelling units, dormitory units, guest rooms and guest suites of hotels and motels, and areas of nursing homes and limited-care facilities used exclusively as patient sleeping rooms, either as an integral part of the service equipment or located immediately adjacent to the service equipment (230.67(A),(B),(C),(D)).

3

Single-Phase Fault Current Analysis

Transformer — 75 VA (utility owned)

Voltage – Primary — 480 volts

   Secondary – 240/120 volts

Percent impedance – Z – 2.6%

$\dfrac{75 \times 1000}{240\ V}$ = 312.50 amperes - Service conductors - 500 kcmil copper-THWN-2-Table 310.16

$\dfrac{100}{2.60 \times 0.90}$ = 42.74

(For transformers that are 25 kVA and larger, UL 1561 recognizes that the actual transformer impedance may be 10% higher or lower than what is marked on the nameplate. This example uses the –10% value to calculate the worst case fault current).

(Transformers built to ANSI standards have a plus or minus 7.5% impedance tolerance.)

$\dfrac{\begin{array}{r}312.50\ \text{amperes}\\ \times 42.74\end{array}}{13,356.25\ \text{amperes}}$ (line – to – line fault current at transformer secondary)

$\dfrac{\begin{array}{r}13.356.25\ \text{amperes}\\ \times 1.50\end{array}}{20,034.38\ \text{amperes}}$ (line - to- neutral fault current at transformer secondary.)

This fault current is higher than the line - to - line fault current at the single - phase center - tapped transformer secondary terminals and, in theory, this value may be $1.33 - 1.67$ times the line - to - line fault current. However, we will use 1.5 in this example.

Service Conductors

$$\frac{500 \text{ kcmil copper-}}{\text{THWN} - 2} \frac{500,000 \text{ cm}}{1973.53} = 253 \text{ mm}^2$$

(conversion factor to determine the equivalent metric wire size)

Service conductor length 25 feet (7.62 meters) - three single conductors (full size neutral) in rigid metal conduit (magnetic). The service conductor(underground) extend to the metering equipment and the service entrance conductor extend from the metering equipment to the service equipment (Article 100) inside the building.

$$\frac{2 \times 25 \text{ feet} \times 13,356.25 \text{ amperes}}{22,185 \text{ "C" value} \times 1 \times 240 \text{ V.}} = \frac{667,812.50}{5,324,400} = 0.1254$$

("C" value table is in Chapter 4)

Note: 22,185 is equal to one over the impedance per foot of the conductor and based on the resistance and reactance values expressed in IEEE Standard 241 (Gray Book) and IEEE Standard 242 (Buff Book)

$$\text{Multiplier M} = \frac{1}{1 + 0.1254} = 0.889$$

$$\begin{array}{r} 13,356.25 \text{ amperes} \\ \times 0.889 \\ \hline 11,874 \text{ amperes} \end{array}$$

(available fault current (L-L) at service equipment (400-ampere panelboard must be marked as suitable for use as service equipment and provided with a two-pole 300 -ampere main circuit breaker (230.66(A)(408.36)), which is the service disconnect (Molded-Case-UL 489). The circuit breaker terminals are identified for conductors at $75\,^\circ\text{C}$ $(110.14(\text{C})(1)(\text{b})(1))$. This panelboard shall be field or factory marked with a warning to qualified persons of potential arc-flash hazards. The marking must meet the requirements in $110.21(B), 110.16(\text{A})$. See the example calculation at the end of Chapter 4 to determine the incident energy label for the service equipment.

Line-To-Neutral Fault Current

$$\frac{2 \times 25 \text{ feet} \times 20,034.38 \text{ amperes}}{22,185 \times 1 \times 120 \text{ Volts.}} = \frac{1,001,719}{2,662,200} = 0.3763$$

Single-Phase Fault Current Analysis 57

$$\text{Multiplier} - M = \frac{1}{1 + 0.3763} = 0.7266$$

20, 034.38 amperes

$$\frac{\times 0.7266}{14, 557 \text{ amperes}}\text{-line-to-neutral fault current at the service equipment.}$$

This establishes the interrupting rating of the 300-ampere circuit breaker (110.9) and the required available fault current marking on the service equipment (110.24(A)).

The service conductors are installed in a 3-inch rigid metal (steel) conduit (3−500 kcmil conductors) (metric designator 78), (Table 4, Chapter 9).

The serving utility has provided a grounding conductor connection from the outdoor transformer neutral point to a grounding electrode as a means of lightning protection and to protect against a possible crossover from primary-to-secondary, as well as protection from transient voltages from their distribution system.

250.24(A)(2) requires at least one additional grounding connection from the grounded service conductor to a grounding electrode elsewhere outside the building as the NEC does not apply to the utility owned transformer. In our example, this will be done from the neutral busbar in the service equipment.

In this example, the Main Bonding Jumper between the neutral bus and equipment grounding bus is in the form of a wire or busbar. So, the grounding electrode conductor may be connected to either the neutral bus or the equipment ground terminal or bus (250.24(A)(4)), within the service equipment. A 6 AWG copper grounding electrode conductor is extended from the neutral terminal bar in the service equipment (250.64(B)(2)) to the grounding electrode outside the building (ground rod).

The grounding electrode in this example is an underground metal water pipe (250.52(A)(1)). Due to the wide use of nonmetallic water piping and the use of this nonmetallic as a replacement for metal water piping when repairs are made, a supplemental electrode is required for the metal underground water pipe (250.53(D)(2)).

This has been a requirement since the 1978 NEC. In this example, the supplemental grounding electrode is a 5/8 inch (15.87mm) diameter by 10 foot (3.048m) copper-clad steel ground rod (250.52(A)(5)) and the ground rod is bonded to the interior metal water pipe within 5 feet (1.52m) from the point of entrance to the building (250.68(C)(1)). The grounding electrode for the interior metal water piping system is 1/0 copper (Table 250.66). The 5-foot length is measured along the water pipe. The upper end of the electrode

58   Single-Phase Fault Current Analysis

is flush with or below ground level, so no additional physical protection is necessary (250.53(A)(4) (250.10)).

The ground rod has been driven to a depth of at least 8 feet (2.44m) in accordance with 250.53(A)(4). And it has been determined that the resistance-to-ground of the ground rod does not exceed 25 ohms, so a supplemental grounding electrode is not required (250.53(A)(2), Exception. A 6 AWG copper conductor is used to connect the ground rod to the metal water pipe (250.66(A)). A listed bolted clamp of cast bronze or brass is used for this connection (250.70(A), (250.8(A)(3)). The 1/0 AWG copper grounding electrode conductors is not exposed to physical damage, so no additional physical protection is necessary (250.64(B)(1)). The 6 AWG copper grounding electrode conductor has physical protection in the form of Schedule 80 PVC 1/2 inch (0.115 square meter (75 mm^2)-Table 4-Chapter 9)(250.64(B)(2)).

Due to the fact that the 1/0 copper grounding electrode conductor for the metal underground water pipe and the 6 AWG copper conductor used to bond the water pipe to the ground rod are installed for the purpose of limiting the imposed voltage associated with lightning and power line surges, typically due to utility switching functions or unintentional contact with higher voltage lines, these conductors should be no longer than necessary to complete their connections. For example, the frequency of lightning ranges from 3 kHz to 10 mHz. So, the increase in inductive reactance due to the increase in frequency will cause a longer conductor to be ineffective in holding systems and equipment at close to earth potential (250.4(A)(1), Informational Note No. 1).

Note that the line-to-neutral fault current is higher than the line-to-line fault current at the single-phase center-tapped transformer secondary terminals and also at the downstream service equipment. This higher fault current will determine the required interrupting rating of the service disconnect, which is a 300-ampere molded-case circuit breaker (110.9). This type of circuit breaker will normally have a means of adjusting the instantaneous-trip-setting. There is a 100-ampere molded-case circuit breaker within the service equipment, which protects a feeder that supplies a 100-ampere, 240/120-V enclosed panelboard. This molded-case circuit breaker will normally have a non-adjustable or fixed instantaneous-trip-setting and fault clearing time of 0.025 seconds (IEEE 1584-Table 1) (molded-case circuit breakers rated less than 1,000 volts with instantaneous integral trip). So the adjustment of the trip-setting of the 300-ampere circuit breaker should assure the selective coordination of these overcurrent devices.

The flexibility that some circuit breakers provide through adjustable pickup settings and time delay features will aid in the selective coordination process, at least in the fault current region. But, the time vs. current curves must be plotted to ensure selectivity in the overload region, as well.

Circuit breakers are not like fuses and the selective coordination process cannot be simply based on the ampere ratings of these devices. For example, a 400-ampere current-limiting circuit breaker that is upstream of a 100-ampere non-current limiting circuit breaker will likely clear before the downstream device. While this may certainly happen with fuses, the available fuse selectivity ratios make it easier to provide selectivity in both the overload and fault current regions.

It is obvious that selective coordination of circuit breakers is dependent on the upstream circuit breaker contacts to remain closed for a long enough period to allow the downstream circuit breaker contacts to open, clear, and isolate the overload or overcurrent condition. Circuit breakers that have higher instantaneous pickup regions and longer short-time delay features should be closer to the source (utility). This is the case in our example, where the 300-ampere molded-case circuit breaker with the adjustable pickup setting is installed ahead of the 100-ampere molded-case circuit breaker that has a fixed instantaneous-trip-setting.

Circuit breakers that are equipped with electronic trip units offer flexibility when it comes to selective coordination and overcurrent protection requirements.

Circuit breaker manufacturers publish circuit breaker to circuit breaker coordination tables that are based on testing. In addition, at least one manufacturer publishes coordination tables based on circuit breaker to fuse coordination, where the circuit breaker is upstream of a current-limiting fuse.

Where selective coordination is applied for fuses, the concept appears to be relatively simple, where the total clearing energy of a downstream fuse is less than the melting energy of the upstream fuse. The available short-circuit current must be calculated and the thermal energy is directly proportional to the square of the current multiplied by the I^2T, which is amperes squared $\times$ time. For example, a 10,000-ampere fault that is cleared in 1/2 cycle (0.008 seconds) will produce a melting energy of 800,000 amperes (10,000 amperes $\times$ 10,000 amperes $\times$ 0.008 seconds = 800,000). The melting energy and the total clearing energy is equal to the melting time plus the arcing time. So, the clearing energy (I^2T) of the downstream fuse must be less than the melting energy (I^2T) of the upstream fuse.

Depending on the size of the electrical distribution system, this may be a tedious process. However, fuse manufactures have simplified this process through the development of fuse selectivity ratio guides. These selectivity

60　*Single-Phase Fault Current Analysis*

ratios normally apply where the available fault current is less than 200,000 amperes or the interrupting rating of the fuses, whichever is less. Using the fuse selectivity ratios, the designer or installer just has to size the fuses properly, based on the connected load and the available fault current.

However, there is more to consider in this process other than just the calculation of the available fault current and sizing the overcurrent protective device in accordance with this available fault current.

A coordination study is critical because the arcing fault current clearing time must be determined by comparing the arcing currents with the time vs. current curves of the overcurrent protective devices in the distribution system. The equations for calculating arcing current are available in IEEE 1584. The total arcing current at each piece of equipment must the determined, as well as the level of the arcing current that will pass through the upstream overcurrent protective device(s). It is the arcing current that passes through the upstream overcurrent protective device(s) that will determine the clearing time of the overcurrent protective device.

Circuit Breakers

There is a simplified approach to identify circuit breaker coordination where the breakers are of the instantaneous-trip type. By multiplying the instantaneous-trip-setting times the circuit breaker ampere rating, the result will be the approximate point where the circuit breaker enters the instantaneous-trip region. This approach is applicable to the instantaneous-trip function, and not to the overload region. In most cases, this method will cover the overload region anyway.

As an example, where a 600-ampere circuit breaker has an instantaneous-trip set at 10 times its ampere rating, or 6,000 amperes, this device will unlatch and open. If this circuit breaker is on the line-side of a 400-ampere circuit breaker that has its instantaneous-trip set at 10 times its ampere rating, or 4,000 amperes, these two devices are selectively coordinated, as long as the available short-circuit current does not equal or exceed 6,000 amperes, where both circuit breakers will open.

This procedure is acceptable, but there is another consideration that relates to a tolerance region. This is a + or – region, which may range from + or – 20% for a thermal magnetic circuit breaker with high-trip-setting, from + or – 25% for a thermal magnetic circuit breaker with a low-trip-setting, or a + or – tolerance of 10% for a circuit breaker with an electronic trip. However, this information should be based on information from the

Single-Phase Fault Current Analysis 61

manufacturer. The positive tolerance may be ignored and only the negative tolerance is considered when using this simple method.

For example, consider an electronic-trip circuit breaker with a rating of 1,000 amperes with an instantaneous-trip set at 10 times and a tolerance of 10% and a downstream 400-ampere thermal magnetic circuit breaker set at 10 times and a tolerance of 25%.

$$(400 \times 10) \times \frac{(1 - 25\%)}{100}$$

$$(4000) \times (1 - 0.25) = 4000 \text{ amperes} \times 0.75 = 3000 \text{ amperes}$$

For an overcurrent of less than 3,000 amperes, the 400-ampere circuit breaker will coordinate with downstream circuit breakers.

For the 1,000-ampere circuit breaker with the instantaneous-trip set at 10 times its rating and a tolerance of 10%, the coordination will be

$$\frac{(1 - 10\%)}{100}$$

$$(100 \times 10)$$

$$(10,000) \times (1 - 0.10) = 10,000 \times 0.9 = 9,000 \text{ amperes}$$

For an overcurrent of less than 9,000 amperes, the 1,000 ampere will coordinate with the downstream 400-ampere circuit breaker. But for short-circuit currents of 9,000 amperes or more, both circuit breakers will open.

Taking these tolerances into consideration, the selective coordination is more accurate.

Circuit breaker manufacturers publish circuit breakers to circuit breaker coordination tables based on testing. Of course, a fault current analysis must be done to calculate the available fault current. Then, based on this calculation, it may be determined if the circuit breakers will coordinate.

Electronic-trip and insulated-case circuit breakers may have a short-time-delay function which allows the circuit breaker a period of time, normally from 6 to 30 cycles (0.096–0.480 seconds), to clear a fault current. This feature is beneficial for low-level fault currents.

However, for higher levels of fault current, there is an instantaneous-trip function, set by the manufacturer at 8−12 times the circuit breaker rating. This instantaneous-trip feature must be considered when determining the selective coordination of the entire system.

62   *Single-Phase Fault Current Analysis*

Equipment Grounding Conductors

The various types of equipment grounding conductors are referenced in 250.118. The minimum sizes of equipment grounding conductors are identified in Table 250.122, based on the ampere rating of the circuit overcurrent device. The actual size of the equipment grounding conductor may, at times, have to be increased due to circuit conditions. These conditions include higher levels of ground-fault current and voltage-drop (250.122(B)). An effective ground-fault current path (250.4(A)(5)),(250.4(B)(4)) must be of sufficiently low impedance, not only to facilitate the operation of the circuit overcurrent device on a solidly grounded system, or to facilitate the operation of the overcurrent devices on an ungrounded or impedance-grounded system in the event of a second ground-fault before the first ground-fault is cleared, but also to limit the voltage-to-ground, above earth potential, on the equipment grounding system. At times, the equipment grounding conductor may be as large as the ungrounded conductor(s). But, in no case, is it required to be larger (250.122(A)).

Another consideration is whether or not to install an equipment grounding conductor that is insulated, covered, or bare (250.118(A)(1)). Due to the fact that 300.3(B) requires all of the conductors of the same circuit, including the equipment grounding conductor, to be installed in the same raceway or cable, or in close proximity in the same trench (300.5(I)) for important impedance reduction, it makes sense that an insulated equipment grounding conductor may be a better choice than one (or more) that is bare. This will reduce thermal stress on adjacent conductors during the ground-fault, until this fault is cleared and, during a ground-fault, if the equipment grounding conductor is not insulated and it is installed in a metallic raceway or cable, there will be arcing within this raceway or cable from the equipment grounding conductor to the metallic enclosure which will damage the insulated conductor within the same raceway or cable. Even for the direct current conductors of the photovoltaic array circuit(s), the equipment grounding conductors must be contained within the same raceway, cable, or otherwise run with these conductors when these circuit conductors leave the vicinity of the PV array (690.43(C)).

But, another important consideration of using insulated equipment grounding conductors for alternating current systems is the reduction of the electromagnetic interference (EMI) on the grounding circuit, as all of the circuit conductors are installed together for important impedance reduction.

After the ground-fault current has been determined by calculation, the actual size of the equipment grounding conductor will be determined.

Single-Phase Fault Current Analysis 63

(Check the UL guide on FHIT cable systems (e.g., cables that are installed on the life/safety and critical branches of an emergency system), where the equipment grounding conductors are installed in raceways and the equipment grounding conductor and raceway is a part of the electrical circuit protection system.)

Beginning with the 1993 NEC cycle, selective coordination was introduced for certain elevator installations. Section 620.62 requires selective coordination for overcurrent devices where more than one driving machine disconnecting means is supplied by the same source. In this case, the overcurrent protective devices in each disconnect must be selectively coordinated with any other supply-side overcurrent devices.

In addition, selective coordination applies to the overcurrent protective devices of Critical Operations Data Systems (645.27), power sources for Electric-motor Driven Fire Pumps (695.3(C)(3)), Emergency System overcurrent devices (700.32), Legally-Required Standby System overcurrent devices (701.32) and Critical Operations Power System overcurrent devices (708.54).

Where Ground-Fault protection is required for the operation of the service and feeder disconnecting means, the feeder GFP must be fully selective with the service GFP so that the downstream device operates to clear a ground-fault on its load side before this fault causes the operation of the service GFP (517.17(C) (Healthcare Facilities) and 708.52(D)) (Critical Operations Power Systems). Of course, the ground-fault protection requirements of 230.95 do not apply to single-phase systems.

Now, returning to our service and feeder example, the fault clearing time of the 100 ampere molded-case circuit breaker should be less than the fault clearing time of the 300 ampere service overcurrent device to assure selectivity, even though this may not be requirement. The downstream 100 ampere panelboard does not require a main circuit breaker, as the upstream 100 ampere circuit breaker in the service equipment provides this protection in accordance with 408.36.

The feeder conductors are installed in a steel (magnetic) raceway (rigid metal conduit). These conductors are 3 AWG copper conductors, THHN, with a full-size neutral conductor. The terminal provisions of the 100 ampere circuit breaker are based on the use of 75°C or higher temperature rated conductors in accordance with 110.14(C)(1)(a)(3)). We have determined that the line-to-line available fault current at the line terminals of the 300-ampere circuit breaker is 11,874 amperes and the line-to-neutral fault current is 14,557 amperes.

64 Single-Phase Fault Current Analysis

The interrupting rating of the 100-ampere two-pole circuit breaker in the service equipment must be no less than 14,557 amperes in accordance with 110.9.

The feeder conductors extend 80 feet (25 meters) to the downstream enclosed panelboard, which is a main lugs only unit. We have established that the fault clearing time of the 100-ampere, two-pole circuit breaker is 0.025 seconds (60 Hz-IEEE 1584-Table 1).

Based on this fault clearing time, we will calculate the available fault current at the downstream panelboard, as well as the insulation withstand rating of the 3 AWG THHN insulated copper conductors:

3 AWG copper conductors – 52,620 circular mils

$\dfrac{52,620 \text{ cm}}{42.25} = 1245$ amperes for 5 seconds (Based on one ampere for every 42.25 circular mils of the conductor cross-sectional area for 5 seconds)

1245 amperes $\times$ 1245 amperes $\times$ 5 seconds $= 7,750,125$ ampere–squared–seconds

$$\frac{7,750,125}{0.025 \text{ seconds}} = 310,005,000$$

$$\sqrt{310,005,000} = 17,607 \text{ amperes}$$

These conductors have an insulation withstand rating of 17,607 amperes for 0.025 seconds, and the available fault current, based on the fault clearing time of the 100-ampere molded-case circuit breaker, is 14,557 amperes. The 3 AWG, THHN copper conductors are properly protected against overload and fault current conditions (110.10).

The available fault current at the 100-ampere panelboard is as follows:

$$\frac{2 \times 80(\text{ feet }) \times 11,874 \text{ amperes}}{4774 (\text{``C'' value }) \times 1 \times 240 \text{ volts}} = \frac{1,899,840}{1,145,760} = 1.66$$

$$m(\text{multiplier}) = \frac{1}{1 + 1.66} = 0.376$$

11,874 amperes

$$\frac{\times 0.376}{4465 \text{ amperes}} (\text{L} - \text{L}) \text{ at panelboard}$$

Single-Phase Fault Current Analysis 65

$$\frac{2 \times 80(\text{feet}) \times 20,034.38 \text{ amperes}}{4774(\text{"C" value}) \times 1 \times 120 \text{ volts}} = \frac{3,205,500.8}{572,880} = 5.595$$

$$m(\text{multiplier}) = \frac{1}{1+5.595} = 0.1516$$

20, 034.38 amperes

$$\frac{\times 0.1516}{3037 \text{ amperes}} (\text{L} - \text{N}) \text{ at panelboard}$$

In this example, there is no main circuit breaker in the downstream panelboard. However, 408.6 requires that this panelboard must have a short-circuit current rating of not less than the available-fault current of 4465 amperes. This fault current (in other than one and two family dwelling units) must be field marked on the enclosure at the point of supply along with the date the calculation was performed. In addition, an arc-flash hazard warning label must be affixed to this panelboard (110.16(A)(B)).

The feeder conductors are installed in a rigid metal conduit (steel) and this conduit serves as the equipment grounding conductor in accordance with 250.118(2). There are 3-3-AWG THHN copper conductors (full-size neutral) installed in a one inch RMC (Table 5–Table 4- Chapter 9). Due to the length of 80 feet and the method of installation, a 1-1/4 RMC may be a better option.

Voltage – Drop

Transformer to Service Equipment

$$\frac{2KXLXI}{CM}$$

K = 12.90 ohms (copper) – 21.20 ohms (aluminum)

L = one way length in feet of conductor

I = amperes of load

cm = circular mil area of conductor (Table 8-Chapter 9)

$$\frac{25.80 \times 25 \text{ feet} \times 312.50 \text{ amperes}}{500,000 \text{ cm}} = 0.4031 \text{volts} \text{ (line–to–neutral)}$$

66 *Single-Phase Fault Current Analysis*

$$\frac{0.4031\,\text{volts}}{2} = 0.20156 \text{ volts} \quad (\text{line–to–line})$$

$$240.00 \text{ volts} - 0.40 \text{ volts} = 239.60 \text{ volts}$$

$$120.00 \text{ volts} - 0.20 \text{ volts} = 119.80 \text{ volts}$$

Service Equipment to Enclosed Panelboard

$$\frac{25.80 \times 80 \times 100 \text{ amperes}}{52,620 \text{ cm}(3 \text{ AWG})} = \frac{206,400}{52,620} \text{ cm} = 3.92 \text{ volts}$$

3.92/2 = 1.96 volts – (line-to-line)

1.96 volts – (line-to-neutral)

$$239.60 \text{ volts} - 3.92 \text{ volts} = 235.68 \text{ volts} \quad (\text{line-to-line})$$

$$119.80 \text{ V} - 1.96 \text{ V} = 117.8 \text{ V} \quad (\text{line-to-neutral})$$

Although the NEC does not specify voltage-drop limits for branch circuit and feeders 210.19- Informational Note and 215.2(A)(2)- Informational Note No. 2 provide guidance for voltage-drop limits.

Also, 647.4(D) identifies the voltage-drop limits for sensitive electronic equipment (120 V. line-to-line and 60 V. line-to-neutral). The voltage-drop is limited to 1.5% for branch circuits and the voltage-drop of feeder and branch circuit conductors is limited to 2.5

The next example is the same as the first example, but the conductors are installed in nonmetallic conduit (Schedule 80 PVC).

Single-Phase Fault Current Analysis

Transformer – 75 kVa – (utility owned)

Voltage – Primary – 480 V

Secondary – 240/120 V

Percent impedance– Z – 2.6%–UL1561 – 2.6% minus 10% – Z at 2.34%

$$\frac{75 \times 1000}{240 \text{ volts}} = 312.50 \text{ ampers}$$

$$\frac{100}{2.60 \times 0.90} = 42.74$$

312.50 ampers

Single-Phase Fault Current Analysis 67

$$\frac{\times 42.74}{13,356.25 \text{ ampers}} = \text{(Line – to line fault current at transformer secondary)}$$

13,356.25 ampers

$$\frac{\times 1.5}{20,034.38 \text{ ampers}} = \text{(Line – to neutral fault current at transformer secondary)}$$

Service Conductors

$500 - \text{kcmil copper } \frac{500,000}{1973.53} = 253 \text{ mm}^2 \text{ (metric wire size)}$

25 feet (7.62 meters) – 3 single conductors in Schedule 80 PVC conduit

13,356.25 L-L-Transformer
20,034.38 L-N-transformer

$$\frac{2 \times 25 \text{ feet} \times 13,356.25}{26,706 \text{ (“C” value)} \times 1 \times 240 \text{ V}} = \frac{667,812.50}{6,409,440} = 0.1042$$

$$\text{m (multiplier)} = \frac{1}{1 + 0.1042} = 0.9056$$

$$\frac{\begin{array}{r}13,356.25 \text{ A} \\ \times 0.9056\end{array}}{12,095.42 A = \text{(L-L) at service equipment}}$$

$$\frac{2 \times 25 \text{ feel} \times 20,034.38}{26,706 \times 1 \times 120 \text{ V}} = \frac{1,001,719}{3,204,720} = 0.3126$$

$$\frac{1}{1 + 0.3126} = 0.7618$$

$$\frac{\begin{array}{r}20,034.38 \text{ amperes} \\ \times 0.7618\end{array}}{15,262.19 \text{ amperes} = \text{(L-N) at service equipment}}$$

The interrupting rating of the 300-ampere molded-case circuit breaker must be no less than 15,262.19 amperes. As in the previous example, this circuit breaker will normally have a means of adjusting the instantaneous-trip-setting. There is a 100-ampere molded-case circuit breaker within the service equipment that protects a feeder that supplies a 100-ampere, 240/120 V panel-board. This molded-case circuit breaker has a fixed instantaneous-trip-setting and a fault clearing time of 0.025 seconds (IEEE 1584-Table 1).

68   *Single-Phase Fault Current Analysis*

The adjustment of the trip-setting of the 300-ampere circuit breaker should assure the selective coordination of these overcurrent devices, at least in the fault current region. But, as in the previous example, especially where selective coordination is a requirement, the time vs. current curves should be examined in order to assure selectivity in the overload region, as well.

These is a 1/0 copper (53.51 mm^2) grounding electrode conductor extended from the neutral terminal bar in the service equipment. And the main bonding jumper is in the form of a wire or busbar, (250.24(A)(4)) as we have already mentioned earlier in this Chapter, a 6AWG copper conductor extends from the neutral bus or the equipment ground bus to a ground rod outside the building (250.24(A)(2)) and, the resistance -to-ground of this ground rod does not exceed 25 ohms (250.53(A)(2), Exception.

The 100 ampere feeder conductors extend a distance of 80 feet (25 meters) to a panelboard. The 100 ampere, 2-pole molded-case circuit breaker protecting the feeder conductors in the service equipment also protects the downstream 100 ampere enclosed panelboard, in accordance with 408.36.

The feeder conductors are 3 AWG THHN copper conductors, including a full-size neutral, and they are installed in a one inch, Schedule 40 PVC conduit (Table 5-Table 4-Chapter 9). The 100 ampere molded-case circuit breaker has termination provisions listed and identified for 75°C in accordance with 110.14(C)(1)(b)(3)). The ambient temperature does not exceed 30°C (86°F).

The equipment grounding conductor, based on the minimum size identified in Table 250.122, is 8 AWG copper. This conductor may be insulated, covered or bare. In our example, this conductor is insulated (250.118(A)(1)) and it is identified with green insulation (210.5(B),(250.119). So, we should be concerned about the ground-fault current, the duration of this ground-fault current in accordance with the operating characteristics of the 100 ampere molded-case circuit breaker protecting the feeder conductors and the insulation withstand rating of the equipment grounding conductor.

The line-to-neutral fault at the service equipment has been calculated at 15,262.19 amperes. So, we will use 50% of this value in our calculation, or 7631.10 amperes.

We have calculated the available fault current at the service equipment at 12,095.42 amperes, line-to-line and 15,262.19 amperes, line-to-neutral. And, we have established the fault clearing time of the 100 ampere molded-case circuit breaker at 0.025 seconds in accordance with IEEE 1584- Table 1 and, this has been determined by examination of the time-current curve characteristics of the 100 ampere circuit breaker.

Single-Phase Fault Current Analysis 69

$$\frac{2 \times 80 \text{ feet} \times 12,095.42 \text{ amperes}}{4811(\text{"C" value}) \times 1 \times 240 \text{ volts}} = \frac{1,935,267.20}{1,154,640} = 1.6761$$

$$\text{m (multiplier)} \frac{1}{1+1.676} = 0.3737$$

$$\begin{array}{r} 12,095.42 \text{ amperes} \\ \times 0.3737 \\ \hline 4520 \text{ amperes} = (\text{L-L at Panelboard}) \end{array}$$

$$\frac{2 \times 80 \text{ feet} \times 15,262.19 \text{ amperes}}{4811(\text{"C" value}) \times 120 \text{ volts}} = \frac{2,441,950.4}{577,320} = 4.23$$

$$\text{m (multipler)} = \frac{1}{1+4.23} = 0.1912$$

$$\begin{array}{r} 15,262.19 \text{ A} \\ \times 0.1912 \\ \hline 2918.13 A = (\text{L-N}) \text{ at panelboard} \end{array}$$

The 100 ampere enclosed panelboard must have a short-circuit withstand rating of no less than 4520 amperes in accordance with 408.6.

Next, we will determine the insulation withstand rating of the 8 AWG copper equipment grounding conductor, based on these conditions.

There are general rules of thumb relating to various fault current types including line-to- line bolted faults, line-to-neutral bolted faults, line-to-ground bolted faults, and line-to- ground arcing faults. These general rules should not replace actual calculations of these faults currents. In our example, we will use this general information to calculate the ground-fault current.

It has been established that 90% of faults occurring in electrical distribution systems are ground-faults. And, 90% of these ground-faults are arcing-type faults. Line-to-ground arcing faults may represent a minimum of 38% of line-to-line bolted faults and line-to- neutral bolted faults. In our example, we will use 50% for our calculations.

Insulation Withstand Rating

8 AWG copper - 50 amperes $-75°C$ (continuous)

8 AWG copper - 16,510 circular mils $- \left(8.37 \text{ mm}^2\right)$

I^2T - ampere - squared seconds

70 *Single-Phase Fault Current Analysis*

I^2T – one ampere for every 42.25 circular mils of conductor cross-sectional area for 5 seconds

$$8 \text{ AWG} \; - \; \frac{16,510 \text{ circular mils}}{42.25} = 391 \text{ amperes} \; - 5 \text{ seconds}$$

$391 \text{ amperes} \times 391 \text{ amperes} \times 5 \text{ seconds} = 764,405$

$$\frac{764,405}{0.025 \; - \; \text{seconds}} = 30,576,200$$

$\sqrt{30,576,200} = 5529 \text{ amperes}$

The insulation withstand rating of the 8 AWG copper conductor for 0.025 seconds is 5529 amperes.

Therefore, if this conductor was subjected to a ground-fault current of 7631.10 amperes for 0.025 seconds, the insulation would be damaged. And, this conductor would not provide the effective ground-fault current path required by 250.4(A)(5). And the protection from extensive damage to electrical equipment of the circuit, as referenced in 110.10 and 110.3(B).

Some people would argue that because 250.118(A)(1) permits the equipment grounding conductor to be bare, the fusing or melting current of this conductor should determine whether or not this conductor provides an effective ground-fault current path.

Fusing or Melting Current

8 AWG copper – 50 amperes –75°C (continuous)

8 AWG copper – 16,510 circular mils – $\left(8.37 \text{ mm}^2\right)$

I^2T - ampere - squared seconds

I^2T – one ampere for every 16.19 circular mils of conductor cross-sectional area for 5 seconds

$$8 \text{ AWG} \; - \; \frac{16,510 \text{ circular mils}}{16.19} = 1020 \text{ amperes} \; -5 \text{ seconds}$$

$1020 \text{ amperes} \times 1020 \text{ amperes} \times 5 \text{ seconds} = 5,202,000$

$$\frac{5,202,000}{0.025 - \text{seconds}} = 208,080,000$$

$\sqrt{208,080,000} = 14{,}425$ amperes

A current flow of 14,425 amperes for 0.025 seconds will cause the 8 AWG copper conductor to reach its melting temperature of 1083°C, after which this conductor will begin to fuse or melt.

So, a potential ground-fault current of 7631.10 amperes for 0.025 seconds will not damage this copper conductor. However, this ground fault current for this duration will probably damage the insulation on the other conductors in the conduit and it is up to the designer/installer and the authority having jurisdiction to make this decision. And, as we have stated in the first example, the protection from EMI on the grounding circuit may also be compromised.

Increasing the equipment grounding conductor from 8 to 6 AWG copper is certainly an important consideration.

6 AWG copper -65 amperes $-75°C$ (continuous)

6 AWG copper $-26,240$ circular mils $- \left(13.30 \text{ mm}^2\right)$

$$6 \text{ AWG} = \frac{26,240 \text{ cm}}{42.25} = 621 \text{ amperes} - 5 \text{ seconds}$$

621 ampere $\times$ 621 amperes $\times$ 5 seconds $= 1,928,205$

$$\frac{1,928,205}{0.025 - \text{seconds}} = 77,128,200$$

$\sqrt{77,128,200} = 8782$ amperes

The insulation withstand rating for the 6 AWG copper conductor for 0.025 seconds is 8782 amperes, which is more than the prospective ground-fault current of 7631.10 amperes.

The fusing or melting current of the 6 AWG copper conductor for 0.025 seconds is as follows:

6 AWG copper $-$ 65 amperes $-$ 75 °C (continuous)

6 AWG copper $-26,240$ circular mils $- \left(13.30 \text{ mm}^2\right)$

$$6 \text{ AWG} = \frac{26.240 \text{ cm}}{16.19} = 1621 \text{ amperes } \text{-5 seconds}$$

72 Single-Phase Fault Current Analysis

1,621 ampere × 1,621 amperes × 5 seconds = 13,138,205

$$\frac{13,138,205}{0.025 - \text{seconds}} = 525,528,200$$

$\sqrt{525,528,200} = 22,924$ amperes

The fusing or melting current of the 6 AWG copper conductor for 0.025 seconds is 22,924 amperes. Once again, this is well above the calculated ground-fault current of 7631.10 amperes and this conductor will not reach its melting temperature.

Based on the use of 3-3AWG- THHN and 1-6 AWG-THHN copper conductors the minimum size of the Schedule 40 PVC conduit will be 1-1/4 inch (374 mm^2) – Table 5 and Table 4 of Chapter 9. (0.581 sq. in.)

Table 5 - Chapter 9

$$3\text{AWG} - \text{THHN} - \frac{0.0973 \text{ sq. in (3AWG – THNN)}}{0.2919 \text{ sq. in}} \times 3 .$$

$$6 \text{ AWG} - \text{THHN} - 0.0507 \text{ sq. in} \quad \frac{0.2919}{\times 0.0507} \\ \overline{0.3426 \text{ sq. in}}$$

Table 4 - Chapter 9

1 1/4 inch conduit –0.5810 sq. in. (metric designator 35)

A 15 ampere branch circuit breaker in the 100 ampere panelboard supplies a row of fluorescent lighting fixtures (luminaires). This lighting load is continuous, so the maximum load on this 15 ampere circuit breaker is 12 amperes, that is 80% of 15 amperes, in accordance with 210.20(A).

The branch circuit conductors are 14 AWG THHN copper, and the total length of these conductors is 80 feet (24.38 meters), including the conductor length through the connected luminaires. There are 10-4 foot (1.219 meters) fluorescent luminaires connected end-to-end 410.64(A),(B),(C). The branch circuit conductors are extended through a 1/2 inch electrical metallic tubing (steel), and the equipment grounding conductor is 14 THHN copper (250.119), (250.118)(A)(1)), (410.42), (410.44). The ballasts are 80% power factor.

Single-Phase Fault Current Analysis 73

And now, the branch circuit supplying the fluorescent lighting load. The total load of the 10 luminaires is 8.33 or 9 amperes.

$$\text{Ballast} - \frac{100 \text{ VA}}{120 \text{ V}} = 0.833 \text{ amperes } (80\% \text{ Power Factor })$$

$$\begin{array}{r} 0.833 \text{ amperes} \\ \times 10 \text{ luminaires} \\ \hline 8.33, \text{ or } 9 \text{ amperes} \end{array}$$

The total lighting load is 9 amperes. The branch circuit conductors are sized at 125% of the lighting load (1.25 × 9 amperes = 11.25 amperes) to satisfy 210.19(A)(1). And, the 14 AWG copper conductor is protected in accordance with 240.4(D)(4) at 15 amperes. There is no temperature correction factor applied, as the ambient temperatures does not exceed 30°C (86°F). And, the circuit breaker terminal provision is for conductors rated at 60°C (110.14(C)(1)(a)(1), or 15 amperes in accordance with Table 310.16.

The voltage-drop at the first luminaire will be as follows:

$$\frac{2 \times 12.9 \times 40 \text{ feet } \times 9 \text{ amperes}}{4,110 \text{ cm}(14 \text{ AWG})} = 2.26 \text{ V}$$

$$\begin{array}{r} 117.84 \text{ V} \\ -2.26 \text{ V} \\ \hline 115.58 \text{ V} \end{array} \text{ (at first luminaire)}$$

Based on these conditions, the line-to-neutral bolted fault current at the first luminaire will be as follows:

$$\frac{2 \times 40 \text{ feet } \times 2918.13 \text{ amperes}}{389(14 \text{ AWG }) \times 1 \times 115.58 \text{ volts}} = \frac{233,450.40}{44,960.62} = 5.1923$$

$$\frac{1}{1 + 5.1923} = 0.1615$$

$$\begin{array}{r} 2918.13 \text{ amperes} \\ \times 0.1615 \\ \hline 471.28 \text{ amperes} \end{array}$$

The voltage-drop at the last luminaire will be as follows:

$$\frac{2 \times 12.9 \times 80 \text{ feet } \times 9 \text{ amperes}}{4110 \text{ cm}} = 4.52 \text{ volts}$$

117.84 volts

74 Single-Phase Fault Current Analysis

$$\frac{-\ 4.52 \text{ volts}}{113.32 \text{ volts (at last luminaire)}}$$

And the line-to-neutral bolted fault current at the last luminaire will be

$$\frac{2 \times 80 \text{ feet} \times 2918.13 \text{ amperes}}{389(14 \text{ AWG} \times 1 \times 113.32} = \frac{466,900.80}{44,081.48} = 10.59$$

$$\frac{1}{1 + 10.59} = 0.0863$$

$$\frac{2918.13 \text{ amperes}}{251.83 \text{ amperes}} \times 0.0863$$

This is significant because, according to the listing instructions for the ballasts of the fluorescent luminaires (410.6),(110.3(B)), these units must be listed to withstand 200 amperes of fault current, and with the calculated fault current for this branch circuit, there is a problem with the provisions of 110.10 and 110.3(B), which states that "listed equipment applied in accordance with their listing shall be considered to meet the requirements of this section" and, "listed, labeled, or both, or identified for a use shall be installed and used in accordance with any instructions included in the listing, labeling, or identification."

These luminaires must have supplementary overcurrent protection (240.10), typically in the form of fuses that are within or external to the luminaire and that provide proper fault current protection for the ballasts (UL1077). These devices are not intended for branch- circuit protection.

The ballasts installed indoors, are also required to have integral thermal protection in accordance with 410.130(E)(1).

We selected a 14 AWG THHN copper conductor for this branch circuit, protected with a 15 ampere circuit breaker (Table 310.16),(240.4(D)(4)). And, based on these circuit conditions, this appears to be acceptable.

However, the line-to-neutral fault current has been determined to be 2918.13 amperes at the 100 ampere panelboard. The 15 ampere molded-case circuit breaker has a fault clearing time of 0.025 seconds. If the line-to-neutral fault developed near this panelboard, will the insulation on the 14 AWG copper conductor be damaged? If so, there is a problem with 110.10 and 110.3(B).

14 AWG copper 4,110 circular mils

$$\frac{4110 \text{ cm}}{42.25} = 97 \text{ amperes for 5 seconds}$$

Single-Phase Fault Current Analysis 75

97 amperes $\times$ 97 amperes $\times$ 5 seconds $= 47,045$ ampere-squared seconds

$$\frac{47,045 \text{ cm}}{0.025 \text{ seconds}} = 1,881,800$$

$\sqrt{1,881,800} = 1372$ amperes for 0.025 seconds

The 14 AWG THHN copper conductor will withstand 1372 amperes for 0.025 seconds without insulation damage. But the available fault current, at least near the panelboard, is 2918.13 amperes and the insulation of the 14 AWG conductor will certainly be damaged under these conditions.

The choice will be to increase the conductor size to increase the insulation withstand rating. And, this increase will be significant. In order to satisfy these conditions, the branch circuit conductors would have be increased to 10 AWG copper (3479 amperes for 0.025 seconds).

12 AWG – THHN copper

$$\frac{6530 \text{ cm}}{42.25} = 155 \text{ amperes}$$

155 amperes $\times$ 155 amperes $\times$ 5 $= 120,125$

$$\frac{120,125}{0.025} = 4,805,000$$

$\sqrt{4,805,000} = 2192$ amperes

10AWG – THHN copper

$$\frac{10,380 \text{ cm}}{42.25} = 246 \text{ amperes}$$

246 amperes $\times$ 246 amperes $\times$ 5 $= 302,580$

$$\frac{302,580}{0.025} = 12,103,200$$

$\sqrt{12,103,200} = 3,479$ amperes

The other solution is to use a current-limiting overcurrent device, either for the branch- circuit or ahead of the branch-circuit, such as the 100 ampere feeder circuit breaker.

Reducing the fault current through the use of current-limiting overcurrent devices for the 100 ampere feeder may be a viable solution. Current-limiting

76   Single-Phase Fault Current Analysis

overcurrent devices, where properly applied will open and clear a fault current in less that 1/2 cycle (0.008 seconds) when these overcurrent devices interrupt current with in their current-limiting range. And, in reducing the available fault current, the arc-flash hazard is significantly reduced, as well.

This is an example where the branch-circuit appeared to be properly protected in accordance with Table 310.16 and 240.4(D)(4). But, there is a problem with 110.10 and 110.3(B), because the short- circuit current rating of the conductor insulation is well below the available fault current.

The 14 AWG THHN copper equipment grounding conductor may be acceptable, based on its fusing current.

14 AWG copper $-4,110$ circular mils

$$\frac{4110 \text{ cm}}{16.19} = 254 \text{ amperes}$$

$$254 \text{ amperes} \times 254 \text{ amperes} \times 5 \text{ seconds} = 322,580$$

$$\frac{322,580}{0.008 \text{ seconds } (1/2 \text{ cycle } 60 \text{ hz})} = 6350 \text{ amperes}$$

If we use the fusing current of the 14 AWG copper equipment grounding conductor, this conductor has a fusing current of 6350 amperes, which is above the fault current of 2918.13 amperes. But, once again, this leads to potential damage to the other circuit conductors with in this raceway and for this reason, this is not acceptable.

So, this analysis includes the pertinent information for the entire distribution system, from the utility transformer to the farthest piece of utilization equipment supplied by a branch-circuit.

4

Three-Phase Fault Current Analysis

The following system analysis comprises an example of a 3-phase distribution system which begins at the secondary of a transformer and extends downstream to the service equipment and then further downstream to a transformer (separately derived system) (250.30).

The secondary of this transformer supplies an enclosed panelboard and a branch-circuit extends to utilization equipment (10hp-3-Phase-208V-squirrel-cage induction motor). This panelboard is in a separate building, in that this building is separated from the adjoining structure by a firewall (Definition of "Building" from Article 100). The supply transformer is **Utility Owned** (90.2(D)(5)), and the transformer secondary connections to the utility owned transformer will establish the service point, that is the point of connection between the facilities of the serving utility and the premises wiring (Article 100). The service conductors (underground) extend to the metering equipment and the service entrance conductors extend from the metering equipment to the service equipment (Article 100) inside the building.

The utility owned transformer will include a connection from the transformer secondary neutral point to a grounding electrode (system). 250.24(A)(2) requires an additional connection to a grounding electrode (system) outside the building, as the NEC and 250.24(A)(2) do not apply to the utility owned transformer (90.2(D)(5)). Outdoor installations are subject to lightning and an accidental crossover from primary to secondary, as well as overvoltages associated with the utility distribution network. So, this additional connection to ground will serve to reduce the overvoltages associated with these conditions. A 6 AWG copper conductor will extend from the neutral busbar within the service equipment to a ground rod outside the building (250.64(B)(2)) this single ground rod has a resistance -to-ground that does not exceed 25 ohms (250,53(A)(2), Exception.

Transformer – 1,000 kVa
Primary – 13,800 V

78 Three-Phase Fault Current Analysis

Secondary – 480/277 V

Transformer %-Z (nameplate) – 3.50%-UL listed transformers that are 25 kVA or larger have a + or – 10% impedance tolerance, in this case 3.85% or 3.15%.

For two-winding transformers built to ANSI standards, the + or – impedance tolerance is 7.5%.

Our calculation will use the worst case of 3.15%Z:

1) The full-load secondary current is

$$\frac{1,000,000 \text{ VA}}{480 \text{ V} \times 1.732} = 1202.848 \text{ or } 1202.85 \text{ A}$$

2) $\dfrac{100}{3.15(Z)} = 31.746$

3) $\begin{array}{r} 1202.85 \text{ amperes} \\ \times 31.746 \\ \hline 38,185.68 \text{ amperes} \end{array}$

In this point-to-point calculation, the short-circuit current that is available at the utility transformer secondary is 38,185.68 amperes. However, there is no motor load included in this calculation. The rotational energy of this type of load will have an effect on the possible fault current produced by this system. Therefore, if the entire load consists of motors, the calculated load current will increase by a factor of 4. In our example, we will assume that 30% of the total load is motor load.

$\begin{array}{r} 1202.85 \text{ amperes} \\ \times 4 \\ \hline 4811.4 \text{ amperes} \\ \times 0.30 \\ \hline 1443.42 \text{ amperes} \end{array}$

$\begin{array}{r} 38,185.68 \text{ amperes} \\ +1443.42 \text{ amperes} \\ \hline 39,629.10 \text{ amperes} \end{array}$

The available fault current at the transformer secondary is now 39,629.10 A.

In our example, we have installed 3-750 kcmil-THWN-2 copper conductors, per phase, with a full-size neutral conductor in Schedule 80 PVC conduit. There are three conduits in parallel (3.10.10(G)(1)). The major portion of

Three-Phase Fault Current Analysis 79

the load is nonlinear, so the neutral conductors are considered to be current–carrying conductors (310.15(E)(3)), (310.15(C)(1)).

The nonmetallic conduits are 5 inch (metric designator 129) Schedule 80 PVC. Table 4–Chapter 9 permits a 4 inch (4.503 square inches-4-750 kcmil-THWN-2 conductors–4.1984 square inches) (metric designator 103), Schedule 80 PVC conduit for this installation, but for wire pulling purposes, the conduit size has been increased to 5 inch.

The length of the secondary conductors from the transformer secondary to the line side of the service disconnect is 80 feet (25 meters).

The service disconnect is a three-pole, molded-case circuit breaker with a rating of 1,200 amperes.

From Table 310.16, the ampacity of the paralleled set of phase conductors is 1605 amperes (3×535 amperes), However, the secondary conductor ampacity, per phase, is 1284 amperes after applying the 80% ampacity adjustment factor from 310.15(C)(1), (310.15(E)(3)). And this conductor ampacity is in compliance with 240.4(C),(240.6(A)), as it is greater than the rating of the 1200 ampere circuit breaker. In addition, the 75°C ampacity of the secondary conductors exceeds the 75°C terminal rating of the 1200 ampere molded-case circuit breaker (110.14(C)(1)(b)(1)).

Assuming a voltage of 480/277V at the transformer secondary and the molded-case circuit breaker size of 1200 amperes, as well as the conductor length of 80 feet (25 meters), the voltage-drop will be as follows:

$$\frac{1.732 \times 12.9 \times 80 \text{ feet} \times 1200 \text{ amperes}}{2,250,000 \text{ cm}} = 0.95329 \text{ or } 1 \text{ volt}$$

$$\begin{array}{r} 480 \text{ volts} \\ -1 \text{ volt} \\ \hline 479 \text{ volts L-L beside volts} \end{array}$$

479 V/1.732 = 276.56L-N

Voltage at the service equipment – 479/276.56 V.

Note: 12.9 = DC resistance for one foot of copper, times the conductor circular mil area. And 21.20 = DC resistance for one foot of aluminum, times the conductor circular mil area. (ANSI/UL 1581-2011- National Bureau of Standards Handbook 100-1966 and 109-1972). Also, see 210.19(A), Informational Note and 215.2(A)(2), Informational Note No. 2 for voltage-drop recommended limits. In addition, see 647.4(D) for voltage-drop limits

80 *Three-Phase Fault Current Analysis*

for Sensitive Electronic Equipment (1.5% for branch circuits and 2.5% for feeders).

Fault-current from Transformer to Service Equipment

$$\frac{1.732 \times 80 \text{ feet} \times 39,629.10 \text{ A}}{29,735(750 \text{ kcmil}) (380 \text{ mm}^2) \times 3(\text{ per phase }) \times 479 \text{ V}} = 0.1285$$

(see Table of "C" values on the next pages)

$$\frac{1}{1 + 0.1285} = 0.8861$$

$$\begin{array}{r} 39,629.10 \text{ A} \\ \times 0.8861 \\ \hline 35,115.35 \text{ A} \end{array}$$ (fault-current at service disconnect)

This fault current establishes the interrupting rating of the 1200 ampere molded-case circuit breaker in accordance with 110.9.

The service conductors are installed in nonmetallic conduits, so there will not be a line-to-equipment ground-fault. However, there may be a bolted line-to-neutral fault which may be as high as 100% of the three-phase bolted-fault near the transformer (39,629.10A), or 50% of this fault current further downstream.

These fault current values must be determined by actual calculations and not by estimates.

In accordance with 110.24(A), the service equipment must be durably and legibly marked in the field with the maximum available fault current. This fault current has been calculated at 35,115.35 amperes, in this example. In addition, the calculation must be documented and made available to those authorized to design, install, inspect, maintain, or operate the system. This field marking is required to be of sufficient durability to withstand the environment (110.24(A)). The service equipment must have a fault current rating that is equal to or greater than the calculated fault current. And, 230.66(A) requires that the service equipment be marked as being suitable for use as service equipment and listed or field evaluated.

In addition, there must be an arc-flash hazard warning that is clearly visible to the qualified persons before examination, adjustment, servicing, or maintenance of this equipment (110.16(A)), (110.16(B)), (110.21(B)). This label must include the following information:

1) Nominal system voltage
2) Available fault current at the service OCPD's

Conductors and Busways "C" Values

AWG or kcmil	Three single conductors						Three conductor cables					
	Steel Conduit			Nonmagnetic Conduit			Steel Conduit			Nonmagnetic Conduit		
	600 V	5 kV	15 kV	600 V	5 kV	15 kV	600 V	5 kV	15 kV	600 V	5 kV	15 kV
Copper												
14	389	–	–	389	–	–	389	–	–	389	–	–
12	617	–	–	617	–	–	617	–	–	617	–	–
10	981	–	–	982	–	–	982	–	–	982	–	–
8	1,557	1,551	–	1,559	1,555	–	1,559	1,557	–	1,560	1,558	–
6	2,425	2,406	2,389	2,430	2,418	2,407	2,431	2,425	2,415	2,433	2,428	2,421
4	3,806	3,751	3,696	3,826	3,789	3,753	3,830	3,812	3,779	3,838	3,823	3,798
3	4,774	4,674	4,577	4,811	4,745	4,679	4,820	4,785	4,726	4,833	4,803	4,762
2	5,907	5,736	5,574	6,044	5,926	5,809	5,989	5,930	5,828	6,087	6,023	5,958
1	7,293	7,029	6,759	7,493	7,307	7,109	7,454	7,365	7,189	7,579	7,507	7,364
1/0	8,925	8,544	7,973	9,317	9,034	8,590	9,210	9,086	8,708	9,473	9,373	9,053
2/0	10,755	10,062	9,390	11,424	10,878	10,319	11,245	11,045	10,500	11,703	11,529	11,053
3/0	12,844	11,804	11,022	13,923	13,048	12,360	13,656	13,333	12,613	14,410	14,119	13,462
4/0	15,082	13,606	12,543	16,673	15,351	14,347	16,392	15,890	14,813	17,483	17,020	16,013
250	16,483	14,925	13,644	18,594	17,121	15,866	18,311	17,851	16,466	19,779	19,352	18,001
300	18,177	16,293	14,769	20,868	18,975	17,409	20,617	20,052	18,319	22,525	21,938	20,163
350	19,704	17,385	15,678	22,737	20,526	18,672	22,646	21,914	19,821	24,904	24,126	21,982
400	20,566	18,235	16,366	24,297	21,786	19,731	24,253	23,372	21,042	26,916	26,044	23,518
500	22,185	19,172	17,492	26,706	23,277	21,330	26,980	25,449	23,,126	30,096	28,712	25,916
600	22,965	20,567	17,962	28,033	25,204	22,097	28,752	27,975	24,897	32,154	31,258	27,766
750	24,137	21,387	18,889	29,735	26,453	23,408	31,051	30,024	26,933	34,605	33,315	29,735
1,000	25,278	22,539	19,923	31,491	28,083	24,887	33,864	32,689	29,320	37,197	35,749	31,959

82 Three-Phase Fault Current Analysis

AWG or kcmil	Three single conductors						Three conductor cables					
	Steel Conduit			Nonmagnetic Conduit			Steel Conduit			Nonmagnetic Conduit		
	600 V	5 kV	15 kV	600 V	5 kV	15 kV	600 V	5 kV	15 kV	600 V	5 kV	15 kV
Aluminum												
14	237	—	—	237	—	—	237	—	—	237	—	—
12	376	—	—	376	—	—	376	—	—	376	—	—
10	599	—	—	599	—	—	599	—	—	599	—	—
8	951	950	—	952	951	—	952	951	—	952	952	—
6	1,481	1,476	1,472	1,482	1,479	1,476	1,482	1,480	1,478	1,482	1,481	1,479
4	2,346	2,333	2,319	2,350	2,344	2,333	2,351	2,347	2,339	2,353	2,350	2,344
3	2,952	2,928	2,904	2,961	2,945	2,929	2,963	2,955	2,941	2,966	2,959	2,949
2	3,713	3,670	3,626	3,730	3,702	3,673	3,734	3,719	3,693	3,740	3,725	3,709
1	4,645	4,575	4,498	4,678	4,632	4,580	4,686	4,664	4,618	4,699	4,682	4,646
1/0	5,777	5,670	5,493	5,838	5,766	5,646	5,852	5,820	5,717	5,876	5,852	5,771
2/0	7,187	6,968	6,733	7,301	7,153	6,986	7,327	7,271	7,109	7,373	7,329	7,202
3/0	8,826	8,467	8,163	9,110	8,851	8,627	9,077	8,981	8,751	9,243	9,164	8,977
4/0	10,741	10,167	9,700	11,174	10,749	10,387	11,185	11,022	10,642	11,409	11,277	10,969
250	12,122	11,460	10,849	12,862	12,343	11,847	12,797	12,636	12,115	13,236	13,106	12,661
300	13,910	13,009	12,193	14,923	14,183	13,492	14,917	14,698	13,973	15,495	15,300	14,659
350	15,484	14,280	13,288	16,813	15,858	14,955	16,795	16,490	15,541	17,635	17,352	16,501
400	16,671	15,355	14,188	18,506	17,321	16,234	18,462	18,064	16,921	19,588	19,244	18,154
500	18,756	16,828	15,657	21,391	19,503	18,315	21,395	20,607	19,314	23,018	22,381	20,978
600	20,093	18,428	16,484	23,451	21,718	19,635	23,633	23,196	21,349	25,708	25,244	23,295
750	21,766	19,685	17,686	25,976	23,702	21,437	26,432	25,790	23,750	29,036	28,262	25,976
1,000	23,478	21,235	19,006	28,779	26,109	23,482	29,865	29,049	26,608	32,938	31,920	29,135

"C" values to conductors

Note: These values are equal to one over the impedance per foot, and based upon resistance value found in IEEE Std. 241-1990 (Gray Book) Recommended Practice for Electric Power Systems in Commercial Buildings IEEE & Std. 242-1986 (Bulf Book), IEEE Recommended Practice for Protection and Coordination of industrial and Commercial Power Systems. Where resistance and reactions values differ or are not available, the Bulf Book values have been used, The values for reactance in determining the C value at 5 kV & 15 kV are from the Gray Book only (Values for 14-10 AWG at 5 kV and 14-B AWG at 15 kV are not available and values for 3 AWG have been recommated.

Three-Phase Fault Current Analysis 83

3) The service equipment OCPD's clearing time, based on the available fault current at the service equipment
4) The date the label was applied

The service conductors from the service point at the transformer secondary connection do not have overcurrent protection. The utility transformer is connected delta-to-wye and there is 30 degree phase shift from primary to secondary. That is, the primary voltage will lead the secondary voltage by 30 degrees. The overcurrent protection on the primary side of this transformer is probably set at 3-6 times, or higher, than the full-load primary current (41.84A).

These overcurrent devices will not provide overcurrent protection for the secondary conductors unless this is a bolted fault within or very close to the transformer. And the service conductors are installed in parallel, so a phase-to-neutral fault in one of the nonmetallic raceways will certainly not produce enough current to open the primary overcurrent protection.

Of course, a primary-to-secondary crossover within the transformer enclosure should open the primary overcurrent protection, but if the transformer is properly installed and maintained, this type of fault is unlikely to occur.

The installation of cable limiters in series with each ungrounded secondary conductor is an important consideration (230.82(1)). These are current-limiting devices and they serve as a means of protection against high fault-currents on the line-side of the service equipment and isolated a faulted cable (s) from the supply system.

Of course, the NEC does not apply to this utility owned equipment (90.2(D)(5)), so Table 450.3(A) does not apply.

However, 230.90 does require that each of the ungrounded service conductors be provided with **overload protection,** which is the 1200 ampere, thermal-magnetic, molded-case circuit breaker within the service equipment. The secondary conductor ampacity is 1605 amperes, per phase, without any correction or adjustment factors applied to the sizing of the primary overcurrent protection.

This satisfies 230.90(A), as the overcurrent protection is in series with each ungrounded service conductor, and this overcurrent protection does not exceed the allowable ampacity of the service conductors.

Selective coordination may be a requirement for this installation. In our example, we will include selective coordination as a requirement (517.31(G)), (620.62), (645.27), (695.3(C)(3)), (700.32), (701.32), (708.54).

84   *Three-Phase Fault Current Analysis*

The 1200 ampere, thermal-magnetic circuit breaker, which is the service disconnecting means, has a maximum adjustable instantaneous trip-setting of 8 times its rating. An easy method of determining the approximate point where the circuit breaker reaches its instantaneous-trip region is to multiply the circuit breaker rating by its trip-setting. In this case, 8×1200, or 9600 amperes. The circuit breaker manufacturer must be consulted to determine the instantaneous-trip pickup tolerance. In our example, we will use a pickup tolerance of + or −20%. So, 1.20×9600 amperes = 11,520 amperes or 0.80×9600 amperes = 7680 amperes. We will use 7680 amperes for selective coordination purposes.

Because this system is solidly grounded, we will use a metal in-ground support structure, which is in direct contact, vertically, with the earth for 10 feet (3.0m) or more, as the grounding electrode. This support structure is bonded to the metal frame of the building, which is also connected to ground through the rebar in the concrete footings (250.52(A)(2)(3)),(250.68(C)(2)). A copper grounding electrode conductor is extended from the neutral busbar in the service equipment to the grounding electrode system. However, if the Main Bonding Jumper (250.28)(A),(B),(C),(D)) is a wire or busbar, the grounding electrode conductor may be connected to the equipment grounding terminal bar (250.24(A)(4)). From Table 250.66, this conductor is 3/0 AWG copper. This conductor is certainly large enough for the supply system, and the most important part of the installation is its length and the method of making the connection to the grounding electrode (system). In order to limit the voltage-rise above earth potential on the connected systems and equipment, the grounding electrode conductor must not be any longer than necessary (250.4(A)(1), Informational Note No. 1).

However, where it may be necessary to extend the grounding electrode conductor for greater lengths, say beyond 100 feet (30.48 meters), there is a method to increase the conductor size to compensate for voltage-drop. For example, where the ungrounded conductor size for a service or separately derived system is 2/0 copper the size of a copper grounding electrode conductor is 4AWG copper according to table 250.66. The short-time rating of the a 4AWG conductor is 988 amperes, based on one ampere for every 42.25 circular mils of the conductor cross-sectional area for 5 seconds (for aluminum conductors it is 64.63 circular mils for 5 seconds).

4 AWG 41.740 cm/42.25=987.93(988) amperes

Table 8-Chapter 9-4 AWG uncoated copper -0.308 ohms/1000 feet

Three-Phase Fault Current Analysis 85

0.308 ohms   988 amperes
×0.10   ×0.10

0.0308 ohms-100 feet 30.43 volts-drop

This is the maximum acceptable voltage-drop for the grounding electrode conductor.

If this conductor were 150 feet long, the conductor size would be increase follows in order to insure that the voltage-drop does not exceed 30.43 volts.

3 AWG-copper-0.245 ohms-1000 feet

0.245   988 amperes
×0.15(150 feet)   ×0.03675

0.03675   36.31 volts-drop

2 AWG-copper -0.194 ohms -1000 feet

0.194 ohms   988 amperes
×0.15(150 feet) ×0.0291 ohms

0.0291 ohms   28.75 volts-drop

Therefore, the grounding electrode conductor is 2 AWG copper, which will limit the voltage-drop to no more than 30.4 volts for the 150 foot length.

The Main Conductor size is dependent on the height of the building. A Class I building has a height of 75 feet (23 m) or less and based on this height, the copper conductor cross sectional area is 57,400 circular mils 29mm^2 (3 AWG-52,620 cm-2AWG-66,360 cm). A class II building has height of over 75 feet (23 m) and the area of the copper cable 105,000 circular mils (1/0 AWG-105,600cm-2/0 AWG-133,100 cm). The construction of this type of cable is somewhat different than normal building wire. The stranding is arranged in a lace pattern in order to make the cable more flexible and to expose as much of the cable cross-sectional area to its surface as possible to reduce the cable impedance due to the skin effect of the alternating current associated with the higher frequency lightning current .This is principle behind the flat conductor construction of a signal reference grid used in Information Technology Equipment rooms.

Aluminum conductors are acceptable where these conductors are not subject to the corrosive effects of galvanic action associated with contact with dissimilar materials, such as copper terminations (110.14). For Class I buildings the aluminum conductor size is 98,600 circular mils (50 mm^2) (1AWG-83,690 cm-1/0-105,600 cm). For Class II buildings, the aluminum conductor size is 192,000 cm (97 mm^2) (3/0 AWG -167,800 circular mils-4/0 AWG -211,600 cm).

86   *Three-Phase Fault Current Analysis*

The connection of the strike termination devices (air terminals) to the grounding electrode system is externally important . This is accomplished through the down conductors, as most of the lightning current will flow to the earth through this conductor.

Other metallic paths which parallel the down conductors will share this flow of current. keep in mind that this current may be on the order of tens or hundreds of thousands of amperes. We have stated in previous references that the duration of this flow of current may range from 2-10 microseconds. So, the fusing or melting current of the Main Conductor and Down Conductors must be a consideration. Installing the down conductors to the grounding electrode system in a manner that will reduce the number of bends in this conductor and where bends are necessary, they must be gradual with a bend radius of no less than 8 inches. (203.2 mm).

There will be at least two down conductors and at least one down conductor for every 100 feet (30.48 meters) of the building or structure perimeter. These conductors will be called upon to carry significant current and there will be an associated voltage-drop with this current flow to the earth. So, the location of these conductors is a primary concern because of the possibility of touch-potential and based on the resistance-to-ground of the grounding electrode system, a step-potential risk, as well.

For example, a ground-potential rise will develop as current flows into the grounding electrode system and this is in accordance with amount of current and the resistance-to-ground system.

Installing a low resistance-to-ground system is important in order to reduce this ground potential rise and to reduce the effects of touch and step potential differences. This is true for any type of system, but especially true for lightning protection system. For this reason, the lightning down conductors and their connection to the grounding electrode system should not be located in areas where people may be present. Probably, the best grounding electrode system for lightning protection is a ground ring (250.52(A)(4)). This type of grounding electrode will serve to reduce the voltage gradient around the building of structure better than other grounding electrodes and it is a common practice to supplement the ground ring with driven ground electrode to lower its resistance-to-ground. And, it is a requirement to bond this grounding electrode for the lightning protection system to the building or structure grounding electrode system

(250.106). And this will serve to reduce the resistance-to-ground of the grounding electrode system and also limit touch and step potential differences.

Three-Phase Fault Current Analysis 87

Where possible, the frame of a steel structured building may serve as a better means of protection than conventional main conductors and down conductors due to its physical size and the increase in the number of lightning down conductors and the connection of the structural steel to a concrete–encased grounding electrode (250.52 (A)(3)).The building steel is not normally accessible to the general public and the concrete-encased electrode typically has a low resistance-to-ground. There have been reports that currents on the order of 500-26000 amperes have caused damage to the concrete footings of buildings, as this current flow through the absorbed moisture in the concrete causes the rapid expansion of this moisture and this resulting damage to the concrete. This condition may be a concern, but to date, neither NFPA 70 or 78 prohibits the use of the concrete-encased electrode and this type of grounding electrode has been in use since 1942 and it is recognized in IEC 62305-3 as an acceptable grounding electrode for lightning protection systems.

And Finally, as we have stressed in other parts of this book, the resistance-to-ground of the grounding electrode system is an important consideration. However, the conductor(s) that is used to make the connection to the grounding electrode (systems) must be limited in length and installed in such a way to be as straight as practicable and unnecessary bends in this conductor should be avoided.

Physical protection for the down conductors to protect against touch potential may be through the use of nonmetallic raceways (Schedule 80 PVC).

The fusing or melting current of the Class I Main Conductor for 5 seconds is as follows:

$$\frac{57,400 \text{ cm}}{16.19} = 3545.40A \text{ (5 seconds)}$$

3545.40A × 3545.4A × 5 seconds = 62,849,305.80A

For one cycle, the fusing current is as follows

$$\frac{62,849,305.80}{0.016 \text{ seconds}} = 3,928,081,612.50$$

$(\sqrt{3,928,081,612.50})= 62,674.41 \text{ A}$

And the fusing current of the Class II Main Conductor for 5 seconds is as follows:

$$\frac{115,000 \text{ cm}}{16.19} = 7103.15A \text{ (5 seconds)}$$

88   Three-Phase Fault Current Analysis

For one cycle, the fusing current is as follows:

7103,15A × 7103,15A × 5 seconds = 252,273,700

$$\frac{252,273,700}{0.016 \text{ seconds}} = 15,767,106,250$$

($\sqrt{15{,}767{,}106{,}250}$)= 125,567.14 amperes

As can be seen here, the short-time (one-cycle) fusing currents for these conductors is very high. And considering that lightning currents reach a peak in 2-10 microseconds, it is very easy to see that these conductors will carry significantly higher currents without fusing or melting.

A radial grounding electrode system consists of one or more main size conductors which extend radially from the lightning down conductor(s) for a length of at least 12 feet (3.7 m) and at least 18 inches (450 mm) below grade.

Once again, the main concern here is not necessarily the buried length of these conductors, but their depth and the soil resistivity with the increase in depth. A rule of thumb that has been in use for many years is that when you double the depth of a ground rode, its resistance-to-ground will be reduced by about 40%. This is due to the normal increase the moisture, possibly mineral content and soil temperature.

Strike termination devices are normally arranged as a series of air terminals that are bonded by the main conductor. These devices are normally installed at least 10 inches (250 mm) above the structure. The spacing between the air terminals will be no more than 20 feet (6 meters). However, with the air terminals are 2 feet (600 mm) or more above the roof, they may be spaced 25 feet (7.62 meters) apart. The air terminals shall also be spaced within 2 feet (600 mm) from the corner of the roof.

There will normally be at least 2 down conductors and this will reduce the voltage-drop as compared to the use of single down conductor. And, at least two down conductors from the strike termination network are required. Where the perimeter of the structure exceeds 250 feet (76 meters), there will be at least 1 down conductor for every 100 feet (30 meters) or less of perimeter length.

Additional down conductors, beyond the minimum, will serve to reduce the voltage-drop and resultant voltage drop of the entire system. However, their location with respect to people and equipment is a primary concern, due to problems associated with touch potential and possible flashovers to nearby metallic equipment.

Three-Phase Fault Current Analysis 89

At one time, it was believed that air terminal must be pointed. In fact, it is said Benjamin Franklin believed that this was important. However, in more recent times, this has been found not to be the case.

And Finally, for equipment installed at ground level, and especially when this equipment is not protected by "rolling sphere method" of protection, the method of grounding and bonding of the structural metal supporting structure, such as that used for the support of PV modules, must be in compliance with NEC 690.43 (A)(B). The devices and systems used for bonding module frames must be listed, labeled and identified for bonding PV modules.

It is a common practice to ground the supporting structure through the use of driven or supporting metal stanchions. This may be an acceptable practice (690.47 (A)), however, from a lightning protection stand point, at least one of these metal supports must extend, vertically, at least 10 feet (3 meters) into the earth in order to comply with 250.52(A)(2) and NFPA 780. In addition, the earth must be compacted and made tight against the length of metal support.

Bends and loops in this conductor should be avoided (250.4(A)(1), Informational Note No.1). The currents produced by lightning have an alternating current component that has a frequency range of 3 kHz to 10 mHz. Frequency is a function of inductive reactance, as $XL = 2\pi FL$. In order to limit the inductive reactance and the overall impedance of the earth connection, the length of this conductor and its method of installation is the primary concern for example,NFPA 780 requires that the bending radius of the Main cable for the connection of the air terminals of lightning protection systems, as well as the lightning down conductors for connections to the grounding electrode system to be a minimum of 8 inches (203.2 mm). This is in recognition of the high frequency of lightning currents the 'skin' effect of this high frequency. The stranding of the Main Conductor is not the same as standard building wire. The flow of electrons will be toward to the surface of the wire and not so much through the core. Therefore, the installation of grounding electrode conductors should follow this principles so that their overall length is limited and any bends in this cable, must be gradual where necessary and not sharp bends.

The other consideration is the size of the Main Bonding Jumper. This conductor will typically be significantly smaller than the ungrounded service conductors. Virtually, all of the ground-fault current will flow through this conductor until the appropriate overcurrent device clears this current, or until the ground-fault relay senses the ground-fault and signals the disconnecting means to open. However, a small percentage of this current will return to the source through other conducting paths, such as through the earth between

90 Three-Phase Fault Current Analysis

the grounding electrode (system) at the service equipment and the grounding electrode (system) at the supply transformer.

250.24(B) states that an unspliced Main Bonding Jumper shall be used to connect the equipment grounding conductors(s) and the service-disconnect enclosure to the grounded conductor in accordance with 250.28. The minimum size of the Main Bonding Jumper is in accordance with Table 250.102(C)(1). In our example, the ungrounded service conductors are 2250 kcmil, per phase, copper, and, in accordance with Note 1 of Table 250.102(C)(1), the copper Main Bonding Jumper minimum size will be 12.5% of the area of the ungrounded phase conductors, or 12.5% of 2250 kcmil.

$$\frac{2,250,000 \text{ cm}}{\times 0.125}$$

281, 250 cm or 300, 000 cm (Table 250.102(C)(1)(Note1)

The Main Bonding Jumper is 300 kcmil copper and the available fault current at the line terminals of the service disconnect has been calculated at 35,115.35 amperes. The minimum phase-to-ground arcing fault may be 13,343.83 amperes (38% of 35,115.35 amperes). But, even if we use the bolted short-circuit current of 35,115.35 amperes, the 300 kcmil copper Main Bonding Jumper will be more than sufficient. In addition, we have already stated that their is a 6AWG (13.30 mm squared)copper conductor extended to an external ground rod and this conductor is provided with physical protection (Schedule 80 PVC -1/2 inch(75 mm squared)) to states 250.64(B)(2).

(See Table of conductor fusing currents)

We have established that the thermal-magnetic molded-case circuit breaker, which is the service disconnect has a maximum adjustable-trip setting of 8 times its rating. And based on the approximate point where this circuit breaker reaches its instantaneous-trip region, minus the 20% pickup tolerance, this circuit breaker will enter its instantaneous-trip region at 7680 amperes. So the 300 kcmil copper Main Bonding Jumper is more than sufficient, even if the fault current is based on the available fault current at the line ter-minals of the service disconnect, as its fusing or melting current, based on the instantaneous-trip region of the thermal-magnetic molded-case circuit breaker at 7680 amperes, is well above this current.

The supply system for this installation is a 3 phase, 4-wire, wye connected service system operating at over 150 volts-to-ground and not exceeding 1000 volts phase-to-phase and the service disconnect is rated 1000 amperes,

Three-Phase Fault Current Analysis 91

FUSING OR MELTING CURRENT

1,083 °C Maximum

AWG	Normal	5 second	1 second	One cycle — 0.016 sec.	1/2 cycle — 0.008 sec.	1/4 cycle — 0.004 sec.	1/8 cycle — 0.002 sec.
	75 °C	1,083 °C	1,083 °C	1,083 °C	1,083 °C	1,083 °C	1,083 °C
14	20 A	254 A	568 A	4,490 A	6,350 A	8,980 A	12,700 A
12	25 A	403 A	901 A	7,124 A	10,075 A	14,248 A	20,150 A
10	35 A	641 A	1,433 A	11,331 A	16,025 A	22,663 A	32,050 A
8	50 A	1,020 A	2,281 A	18,031 A	25,500 A	36,062 A	51,000 A
6	65 A	1,621 A	3,625 A	28,656 A	40,525 A	57,311 A	81,050 A
4	85 A	2,578 A	5,765 A	45,573 A	64,450 A	91,146 A	128,900 A
3	100 A	3,312 A	7,406 A	58,548 A	82,800 A	117,097 A	165,000 A
2	115 A	4,101 A	9,170 A	72,461 A	102,475 A	144,922 A	204,950 A
1	130 A	5,169 A	11,558 A	91,376 A	129,225 A	183,105 A	258,950 A
1/0	150 A	6,523 A	14,586 A	115,311 A	163,075 A	230,623 A	326,150 A
2/0	175 A	8,221 A	18,383 A	145,328 A	205,525 A	290,656 A	411,050 A
3/0	200 A	10,364 A	23,175 A	183,211 A	259,100 A	366,423 A	518,200 A
4/0	230 A	13,070 A	29,225 A	231,047 A	326,750 A	462,094 A	653,500 A
250 kcmil	255 A	15,442 A	34,529 A	272,979 A	386,050 A	545,957 A	772,100 A
300 kcmil	285 A	18,530 A	41,434 A	327,567 A	463,250 A	655,134 A	925,500 A
350 kcmil	310 A	21,618 A	48,339 A	382,156 A	540,450 A	764,312 A	1,080,900 A
400 kcmil	335 A	24,707 A	55,247 A	436,762 A	617,675 A	873,524 A	1,235,350 A
500 kcmil	380 A	30,883 A	69,056 A	545,939 A	772,075 A	1,091,879 A	1,544,150 A

or more. So, ground-fault protection of equipment is required for this installation.

Beginning with the 1971 NEC cycle, ground-fault protection of equipment became a part of Article 230, specifically in 230.95. This protection applied to solidly grounded, 3 phase, 4 wire, wye connected systems, operating at up to 600 volts, phase-to-phase, and, over 150 volts, phase-to-neutral, where the rating of the service disconnect was 1000 amperes, or more. Now, this protection applies to voltages of more than 150 volts-to- ground, but not exceeding 1000 volts phase-to-phase for each service disconnect rated 1000 amperes or more. Since 1971, this type of protection has been expanded to include feeders (215.10), the building or structure main disconnecting means (240.13), 517.17 for Health Care Facilities and 708.52 for Critical Operations Power Systems, where two levels of ground-fault protection are required, and 700.6(D) for Emergency Systems, as well as 701.6(D) for Legally-Required Standby Systems. However, for these last two systems, there will be a ground-fault sensor, located at, or ahead of, the main disconnecting means. And the maximum setting of the signal devices shall be for a ground-fault current of 1200 amperes (700.6(D)), (701.6(D)). In these cases, it is not required to have automatic ground-fault protection, but a means of identifying a ground-fault, so that it may be cleared when the normal service is restored (700.31 and 701.31). Instructions on the course of action to be taken in the event of an indicated ground-fault must be located at or near the sensor location.

92 *Three-Phase Fault Current Analysis*

Of course, where ground-fault protection is required, selective coordination must be considered. In fact, it may be better to assure the selective coordination of the downstream overcurrent devices before considering the settings of the GFP. These settings include the pick-up setting associated with the ampere setting and the time delay setting. This equipment only recognizes ground-faults and provides no protection against short-circuits (phase-to-phase).

The current limit referenced in 230.95 is 1200 amperes, with a maximum time-delay of one second for ground-faults of 3000 amperes, or higher. The ground-fault protective relay trip-setting must be selectively coordinated with the downstream overcurrent devices to eliminate the possibility of a total system "blackout". Unfortunately, when this has happened, sometimes due to the fact that the ampere setting, as well as the instantaneous-trip setting, are too low, this protection is compromised and the ground-fault protection is lost, as, in some cases, the GFP is disconnected.

If it is a problem to selectively coordinate the ground-fault protection with the downstream overcurrent devices, it may necessitate the use of multiple smaller overcurrent devices in lieu of one large feeder or service overcurrent device. For example, 5-800 ampere overcurrent devices, as opposed to 1-4000 ampere overcurrent device, or, the use of an ungrounded or impedance-grounded system(s), with a solidly-grounded system located downstream from the source (230.71),(250.4(B)),(250.36),(250.187).

For healthcare facilities and critical operations power systems, there must be two levels of ground-fault protection (517.17(B) and 708.52(B)). And these two levels of protection must be fully selective. At one time, 517.17 specified at least a six-cycle separation between the upstream ground-fault protection relay time band and the downstream ground-fault relay. Now, 517.17(C) and 708.52(D) require that the downstream (feeder) ground-fault protection opens on ground-faults on the load side of the feeder device without affecting the ground-fault protection at the service. Separation of the ground-fault time–current characteristics shall be in conformance with the manufacturer's recommendations to achieve 100% selectivity. In any event, the ground-fault protection must be tested when first installed to ensure that each level of protection is operational (517.17(D),(708.52(C)).

Selective coordination is a requirement for Emergency Systems (700.32) and legally required Standby Systems (701.32). The selective coordination is selected by a licensed professional engineer or other qualified person(s) engaged in the design, installation or maintenance of electrical systems.

Three-Phase Fault Current Analysis 93

Designing the distribution system to be fully selective, including where ground-fault protection is required, can be quite difficult. This is especially true where ground-fault protection of equipment is required. Short-circuit current calculations must identify the available short-circuit currents throughout the distribution system. This will identify the required interrupting ratings of overcurrent devices (110.9), as well as the required operating characteristics of the overcurrent devices installed at various locations, even at the farthest points from the source.

However, there is more to consider in this process, other than just the calculation of the available fault current and sizing the overcurrent protective device in accordance with this available fault current.

A coordination study is critical because the arcing fault current clearing time must be determined by comparing the arcing currents with the time vs. current curves of the overcurrent protective devices in the distribution system. The equations for calculating arcing current are available in IEEE 1584. The total arcing current at each piece of equipment must the determined, as well as the level of the arcing current that will pass through the upstream overcurrent protective device(s). It is the arcing current that passes through the upstream overcurrent protective device(s) that will determine the clearing time of the overcurrent protective device.

However, in considering ground-fault conditions, it must be recognized that, by far, the vast majority of ground-faults are arcing type faults. That is, a strike and restrike between an ungrounded conductor and the equipment grounding system. This may be caused by damage to conductor insulation or by contaminants absorbed through the insulation, possibly dirt combined with moisture, that form tracking paths through the insulation, and possibly, to the equipment grounding system. There is a voltage-drop in this arcing fault, possibly 50 volts, or more, and this compounds the problem in returning the ground-fault current to the source. If not promptly cleared in a system that is solidly grounded, the arcing fault may lead to equipment damage, or possibly result in a short-circuit fault, which may cause more severe damage. And, the possible touch-potential problem for people makes the prompt clearing of this fault even more important.

In an ungrounded system, this fault, which is recognized and identified by the ground detectors required by 250.21(B)(1)(2), must be located and corrected before the arcing ground-fault causes damage to an opposite phase, resulting in a significant short-circuit fault and more damage.

The arcing ground-fault **may** be as low as 38% of a three-phase bolted short-circuit current. Therefore, the equipment grounding system must be

94 Three-Phase Fault Current Analysis

designed to provide an effective ground-fault current path that assures a low-impedance circuit and that will facilitate the operation of the overcurrent device or initiate the operation of the ground detector for an ungrounded or impedance-grounded system (250.4(A)(5),(250.4(B)(4). The effective ground-fault current path must have sufficient current-carrying capacity to safely carry the ground-fault current that may be imposed on it (solidly grounded system), or to promptly clear the overcurrent devices in the event of a second ground-fault on an opposite phase in an ungrounded or impedance-grounded system.

Similar to adjustable-trip circuit breakers, the trip-setting of the GFPE is initially set to the minimum and this leads to many nuisance trips. The trip-setting of the Ground-Fault Protection Relay must be coordinated with the overcurrent devices that actually provide the overcurrent protection. This means that the overcurrent devices must be able to recognize all levels of ground-fault current, regardless of how high or low this current may be.

The ground-fault protection and the downstream overcurrent protection combination has a ground-fault **effective** curve, which is a composite of the GFPE and the OCPD curves. When analyzing line-to-ground faults, the effective GFPE and OCPD curve must be examined.

Certainly, the GFPE must be included in the selective coordination analysis to such an extent that where this type of protection causes a lack of selective coordination, an alteration in the system design may be in order. As we have stated previously, this may include the installation of multiple service disconnects (230.71(B)), or the use of an impedance grounded system (250.36),(250.187), or an ungrounded system with a solidly grounded system(s) downstream.

The grounding electrode that is used in our example is a metal in-ground support structure which is in direct contact, vertically, with the earth for 10 feet (3.0m) or more (250.52(A)(2)).

This support structure is bonded to the metal frame of the building, which is also connected to ground through the tie-down bolts that are connected to a concrete-encased electrode (250.52(A)(3)). In this way, the metal building frame may serve as a conductor to interconnect grounding electrodes that are a part of the grounding electrode system or as a grounding electrode conductor (250.68(C)(2)). In addition, equipment that is secured to grounded metal supports (metal building frame) is permitted to be considered as being connected to an equipment grounding conductor, where this equipment is connected to an equipment grounding conductor (250.134),(250.136). And, we have provided an external grounding electrode (ground rod) in accordance with 250.24(A)(2).

The service phase conductors are 2250 kcmil copper, so Table 250.66 requires a 3/0 copper grounding electrode conductor. However, the key element here is the length and the method of installation of this conductor, including its termination method. The GEC is to be connected to earth in such a way that will serve to limit the voltage associated with lightning, line surges, or unintentional contact with higher voltage lines. The most important factor to consider is the high frequency effects of lightning. The alternating current component of lightning has a frequency that ranges from 3 kHz to 10 mHz. The inductive reactance of alternating current is affected by frequency, as $XL = 2\pi FL$. So due to the higher inductive reactance and overall impedance of the connection to earth, the grounding electrode conductor must be of limited length and the completed connection to the grounding electrode must be made by exothermic welding, listed lugs, listed pressure connectors, listed clamps, or other listed means (250.4(A)(1)),(250.8(A)),(250.70). In addition, the length of this conductor does not exceed 100 feet (30.48 m), so the voltage-drop will not be excessive and increasing the size of this conductor is not necessary.

In this example, the 3/0 AWG-copper grounding electrode conductor is not subject to physical damage, so physical protection is not necessary (250.64(B)(1)). Also, in our example, a concrete-encased electrode is a part of the grounding electrode system (250.52(A)(3)).

The most common type of fault in an electrical distribution system is an arcing ground-fault, which **may** be a minimum of 38% of three-phase bolted fault currents. But, near to the source, the arcing ground-fault current will be significantly higher.

The following Table identifies various fault types and the percentages of three-phase bolted fault currents that each fault type represents. Bolted fault currents are certainly not common in electrical distribution systems.

Fault types	Percentage
Three-phase bolted fault	100%
Line-to-line bolted fault	87%
Line-to-ground bolted fault	25-125% (use 100%) near transformer, 50% further downstream)
Line-to-neutral bolted fault	25-125% (use 100% near transformer – 50% further downstream)
Three-phase arcing fault	89% maximum
Line-to-line arcing fault	74% maximum
Line-to-ground acing fault	38% minimum

96 Three-Phase Fault Current Analysis

Manufacturers use bolted fault currents to test the interrupting ratings of overcurrent devices and the bracing of busbars in distribution equipment and this testing is done in controlled laboratory environments.

This information is general in nature and it is not meant to take the place of the actual calculations provided by software applications. The NEC permits the maximum setting of the GFPE at 1200 amperes and a time-delay of up to one second for currents equal to or greater than 3000 amperes (230.95(A)). Arcing ground-faults are the most common and also the most damaging, as the damage is directly proportional to the amount of time that this fault is allowed to continue.

IEEE 1584-2002 is the source for information relating to the calculations of arcing currents. This standard is the "IEEE Guide for Performing Arc-Flash Hazard Calculations". While the line-to-ground arcing fault current may represent 38% of 3-phase bolted fault currents, this arcing fault current must be calculated, as this arcing current will be used in the calculation of the clearing time of the upstream overcurrent protective device.

A 3-phase, 150 kVa, 480/208/120 volt transformer is supplied from the service equipment and the marked impedance is identified at 1.20%..

250.134(1) normally requires equipment that is fastened in place or connected by permanent wiring methods to be grounded by any of the equipment grounding conductors that are permitted by 250.118. This includes the connection to an equipment grounding conductor that is within the same raceway, cable, or otherwise run with the circuit conductors for important impedance reduction (300.3(B)). As the space between conductors increases, magnetic flux density decreases and the circuit impedance increases. This increase in impedance will reduce the ground-fault current through the equip-ment grounding conductor, causing the circuit overcurrent device to delay its operation, or not clear at all. An effective ground-fault current path is required to be a low-impedance circuit that facilitates the operation of the circuit overcurrent device and limits the voltage-rise on the metal frames of equipment as the ground-fault is cleared. For this reason, 300.3(B) and 300.5(I) require that all of the conductors of the same circuit, including the grounded conductor and all of the equipment grounding and bonding conductors to be contained within the same raceway, auxiliary gutter, cable tray, cablebus assembly, track, cable, or cord, or buried in close proximity in the same trench, unless otherwise permitted in 300.3(B)(1) through (B)(4). On an ungrounded or impedance grounded system, the effective ground-fault current path is meant to assure the prompt clearing of the overcurrent devices

Three-Phase Fault Current Analysis   97

in the event of a second ground-fault from an opposite phase of the wiring system (250.4(A)(5),(B)(4)).

Even though the equipment grounding conductor(s) is permitted to be bare, where the circuit conductors are installed within the same raceway, cable, cord, etc., to comply with 300.3(B), the ground-fault current flowing through this conductor may produce a temperature in this conductor which may damage the other conductor insulation in the same raceway. In addition, where the bare equipment grounding conductor is run with the other circuit conductors, with in a metal raceway, there will be a voltage difference between this conductor and the metal raceway, as this raceway will carry more of the ground-fault current leading to internal arcing between the equipment grounding conductor and raceway, which will likely damage the other circuit conductors. Using an insulated equipment grounding conductor will protect the other conductors from the thermal stress of the ground-fault current. But, another important consideration of using insulated equipment grounding conductors for alternating current systems is the reduction of the electromagnetic interference (EMI) on the grounding circuit, as all of the circuit conductors are installed together for important impedance reduction.

For a direct current system, the equipment grounding conductor may be run separately from the circuit conductors. In this case, the resistance of this conductor is the important consideration, as the circuit impedance is not a factor (250.134, Exception No. 2). However, 690.43(C) permits equipment grounding conductors for photovoltaic array circuits and support structures to be run separately from the PV system conductors within the array. And, where the PV system conductors leave the array, the equipment grounding conductor should comply with 250.134, which means the equipment grounding conductor is to be run with the other PV source conductors.

The load supplied by the transformer primary feeder conductors has been determined to be 120 amperes per phase of continuous load, plus 30 amperes of non-continuous load (215.2(A)(1)). So, the minimum ampacity of the feeder conductors will be 180 amperes (120 amperes $\times$ 1.25 = 150 amperes plus 30 amperes). There is no correction factor to be applied in this example for ambient temperature.

The voltage at the service equipment has been determined to be 479/276.56V.

The 150 kVA transformer has a full-load primary current of 180.43 amperes where the primary voltage is 480 volts. However, this transformer is supplied

98 *Three-Phase Fault Current Analysis*

by a feeder that is 95 feet long. And, as will be determined later, this voltage will be 476.74/275.25V.

$$\frac{150 \text{ kVA}}{476.74 \text{ V}} = \frac{150,000 \text{ kVA}}{476.74 \text{ V} \times 1.732} = \frac{150,000}{825.71} = 181.66 \text{ amperes}$$

Because this transformer will be provided with primary and secondary overcurrent protection (240.4(F)), Table 450.3(B) permits the primary overcurrent protection to be 250% of the full-load primary current. If this is done, these feeder conductors will be sized accordingly. That is, 2.5 times 180.43 amperes, or 451 amperes (240.3),(240.4).

However, in our example, the primary overcurrent device will be rated at 125% of the full-load primary current, or 227.075 amperes (181.66 amperes × 1.25 = 227.075A). Table 450.3(B), Note 1 permits the overcurrent device to be the next standard size, or 250 amperes, but we will use a 225 ampere-3-phase molded-case circuit breaker (750 degree C terminals).

So, the transformer primary conductors will be 4/0 THHN copper, that is 260 amperes at 90 degree C and 230 amperes at 75 degree C, which is the 225 ampere circuit breaker termination provisions (110.14(C)(1)(b)(1)) and 240.4. These conductors are installed in electrical metallic tubing (steel). There will be a 4 AWG THHN copper equipment grounding conductor run with the feeder conductors as a supplement to the EMT (250.118(A)(1)(4)), (Table 250.122).

Table 5 - Chapter 9

4/0 AWG-THHN-0.3237 square inches × 3 = 0.9711 sq. inches

4 AWG-THHN-0.0824 square inches × 1 = 0.0824 sq. inches
1.0535 sq. inches

Table 4 - Chapter 9

Trade size-2 inch EMT-1.342(40%)-Metric Designator 53

There is a six foot length of flexible metal conduit (steel)-(2 inch-40% fill = 1.307 square inches = Table 4 - Chapter 9) connected from the EMT to the primary side of the 150 kVA transformer. The listed flexible metal conduit is not an equipment grounding conductor because of its size and the overcurrent protection for the transformer primary conductors exceeds 20 amperes (250.118(A)(5)(b)(c)) (ANSI-UL1).

A 4 AWG copper bonding conductor may be installed on the outside of the flexible metal conduit where the length is limited to 6 feet

Three-Phase Fault Current Analysis 99

(250.102(E)(2)). However, there is a 4 AWG copper equipment grounding conductor run with the feeder conductors, so no additional bonding is necessary.

There will be a terminal bar for all grounding and bonding connections secured inside the transformer enclosure, as the metal enclosure of the transformer is not listed as a bonding means (450.10(A)).

This dry-type transformer exceeds 112.5 kVA and 450.21(B) requires this equipment to be installed in a transformer room of fire-resistant construction, unless the transformer has a Class 155 or higher insulation system and completely enclosed except for ventilating openings (450.21(B)), Exception No. 2.

Voltage-Drop

$$\frac{1.732 \times 12.9 \times 95 \text{ feet} \times 225 \text{ amperes}}{211,600 \text{ cm}(4/0 \text{ AWG})} = \frac{477,577.35}{211,600 \text{ cm}(4/0 \text{ AWG})} =$$

2.2570 volts or 2.26 volts

$$\frac{479.00 \text{ V.}}{476.74 \text{ V.}} \text{ (line } - \text{ to } - \text{ line)}$$

$$\frac{476.74}{1.732(\sqrt{3})} = 275.25 \text{ volts (line - to - neutral)}$$

$$\frac{476.74 \text{ V.(line } - \text{ to } - \text{ line)}}{275.25 \text{ V.(line } - \text{ to } - \text{ neutral)}}$$

This voltage is within the 3% voltage-drop limit for feeders in accordance with 215.2(A)(2)-Informational Note No. 2. This is not a mandatory requirement (90.5(D)), but this will provide for reasonable efficiency of operation, in most cases.

The feeder conductors are supplying the 150 kVA, 480/208/120 V transformer (1.20%- Z). The available fault current at the transformer primary is as follows:

$$\frac{1.732 \times 95 \text{ feet} \times 35,115.35}{15,082(\text{``C'' value }) \times 1 \times 476.74 \text{ V}} = \frac{5,777,879.69}{7,190,192.68} = 0.8036$$

$$\frac{1}{1+0.8036} = 0.5544$$

$$\frac{35,115.35 \text{ amperes}}{\times 0.5544}$$

$$\overline{19,468 \text{ amperes}}$$

100 *Three-Phase Fault Current Analysis*

The next step is to determine whether this installation is in compliance with 240.92(C). The transformer secondary conductors are a part of a separately derived system (250.30) and there is no overcurrent device where these conductors are attached to the transformer secondary terminals. If we use the first condition referenced in 240.92(C)(1), we can see that the secondary conductors are limited to a length of 100 feet. In our example, the secondary conductor length is 75 feet. The second provision is associated with the rating of the primary overcurrent device, which is the 225 ampere molded-case circuit breaker within the service equipment. The ampere rating of this circuit breaker cannot exceed 150% of the value determined by multiplying the secondary conductor ampacity by the secondary-to-primary transformer voltage ratio.

If the transformer secondary conductors are rated at 75°C, they may be 600 kcmil THHN copper conductors, which have an ampacity of 475 amperes according to Table 310.16 and 110.14(C)(1)(b)(1).

The transformer primary voltage is:

476.74 Volts-Line-to-Line

275.25 Volts-Line-to-Neutral

$$\begin{array}{r} 476.74 \text{ V} \\ \times 0.433 \\ \hline \end{array}$$

$206.43 \text{ V} - \text{Line} - \text{to} - \text{Line (Secondary)}$

$\dfrac{206.43}{1.732} = 119 \text{ V} - \text{to Neutral (Secondary)}$

$\dfrac{150,000 \text{ VA}}{206.43 \times 1.732} = 4.19.53 \text{ amperes (Full} - \text{Load Secondary Current)}$

Let's assume the use of 600 kcmil THHN copper conductors at 475 amperes.

The secondary-to-primary transformer voltage ratio is:

$\dfrac{206.43 \text{ V}}{476.74} = 0.433 \text{ amperes}$

$$\begin{array}{r} 475 \text{ amperes} \\ \times 0.433 \\ \hline 205.68 \text{ amperes} \end{array}$$

$$\begin{array}{r} \times 1.5 (150\%) \\ \hline 308.52 \text{ amperes} \end{array}$$

Three-Phase Fault Current Analysis 101

So, based on these conditions, the 225 ampere transformer primary overcurrent device is in compliance with 240.92(C)(1)(1), as its rating does not exceed 308.52 amperes.

However, 240.4(F) does not permit the 225 ampere transformer primary overcurrent device to protect the transformer secondary conductor, due to the 30 degree phase-shift of this delta-to-wye connected transformer, and Table 450.3(B) permits the transformer secondary overcurrent device to be rated at 125% of the transformer full-load secondary current, or 419.53 amperes × 1.25 = 524.41 amperes. And, Note 1 of Table 450.3(B) permits the use of the next standard size of overcurrent device, or 600 amperes from Table 240.6(A). However, if a 600 ampere circuit breaker or 600 ampere fuses are used, the conductors connected to these overcurrent devices must be protected against overcurrent in accordance with their ampacities (240.3),(240.4),(240.21). So, it would be more appropriate to use a 400 ampere overcurrent device(s), in this case, a molded-case circuit breaker, to protect the 600 kcmil, THHN copper conductors and this is in compliance with the 75°C circuit breaker terminals (110.14(C)(1)(b)(1)). And the use of a 600 kcmil conductor on the terminals of the 400 ampere circuit breaker does not violate the two wire size limit restriction of UL 489 for the molded-case circuit breaker. So the time/current curve characteristics of this overcurrent device will not be affected by the use of the 600 kcmil conductors.

In this example, the major portion of the secondary load in the 3-phase, 4-wire wye connected system is nonlinear. So, the secondary neutral conductor is considered to be a current-carrying conductor (310.15(E)(3)), and the provisions of 310.15(C)(1) will apply. The load on the secondary conductors will be limited to 80% of the normal ampacity of the 600 kcmil-THHN-copper conductors, or 80% of 475 amperes = 380 amperes. A full- size neutral conductor is installed (4-600 kcmil-THHN copper conductors). There will also be a 3 AWG-THHN copper equipment grounding conductor, based on the rating of the 400 ampere circuit breaker (Table 250.122)

The available fault current at the transformer secondary will be

$$\frac{19,468 \text{ A} \times 476.74 \times 1.732 \times 1.2 \text{ Z} \times 0.9}{100,000 \times 150 \text{ kVA}} = \frac{17,360,993.40}{15,000,000} = 1.1574$$

$$\frac{1}{1+1.1574} = 0.4635$$

$$\frac{476.74 \text{ V} \times 0.4635 \times 19,468}{206.43 \text{ V}} = \frac{4,301,824.30}{206.43 \text{ V}} = 20,839.143 \text{ amperes}$$

102 *Three-Phase Fault Current Analysis*

The secondary conductors are installed in a 6 foot (1.829 meter) length of flexible metal conduit to the secondary 400 ampere molded-case circuit breaker enclosure. This enclosure must be listed for its intended use and it must have the wire bending space at terminals and minimum gutter space provided in switch enclosures as required in 312.6(A)(404.3(A)(B)(C)). For the 600 kcmil secondary conductors and one wire per terminal, the minimum wire bending space is 8 inches (203mm). This wire bending space at the terminals of the 400 ampere circuit breaker is measured in a straight line from the end of the lug in the direction that the wire leaves the terminal to the wall of the circuit breaker enclosure.

The transformer secondary is a separately-derived system in accordance with the definition in Article 100 and 250.30(A). A System Bonding Jumper is installed from the neutral point of the wye connected secondary to the terminal bar inside the transformer enclosure (250.12) for the grounding and bonding connections (450.10(A)). The System Bonding Jumper is 1/0 copper in accordance with Table 250.102(C)(1), based on the 600 kcmil copper secondary conductors. The Grounding Electrode conductor is extended from the terminal bar within the transformer enclosure where the System Bonding Jumper is connected (250.30(A)(5)) to the grounding electrode, which is the metal structural frame of the building (250.68(C)(2)). The grounding electrode conductor is 1/0 copper (Table 250.66). And, just as in the case of the grounding electrode conductor at the service equipment, the GEC must not be any longer than necessary to complete the connection to the metal frame of the building in compliance with 250.70(A), 250.8(A), 250.10. Remember, the structural metal frame of the building in this case is a grounding electrode because it extends into the earth for 10 feet or more (250.52(A)(2)). And the metal structural frame of the building is permitted to be used as a grounding electrode conductor (250.68(C)(2).

The grounded conductor of the separately derived system must be bonded to the nearest available point of the metal water piping system in the area served by the separately- derived system (250.104(D)(1)). This bonding conductor will be 1/0 AWG copper. A separate bonding connection to the exposed structural metal frame of this building is not required from the grounded conductor of the separately-derived system because the building structural metal frame is used as the grounding electrode for the separately-derived system (250.104(D)(2) Exception No. 1).

The size of the 6 foot length of the flexible metal conduit for 4-600 kcmil THHN copper phase and neutral conductors and a 3 AWG THHN equipment grounding conductor (Table 5, Chapter 9) 600 kcmil THHN-0.8676 sq.in.

Three-Phase Fault Current Analysis 103

(559.7 mm2)-3 AWG THHN-0.0973 sq. in.

$$\begin{array}{r} 0.8676 \\ \times 4 \\ \hline 3.4704 \text{ sq. in} \\ +0.0973 \text{ sq. in} \\ \hline 3.5677 \text{ sq. in} \end{array}$$

Table 4, Chapter 9 – 3.5 inch (metric designator 91) (3.848 sq. in.)

In addition to the equipment grounding conductor that is run with the ungrounded secondary conductors to the secondary overcurrent device, there is a 3 AWG copper bonding jumper on the outside of the flexible metal conduit. This bonding jumper is not required, due to the internal equipment grounding conductor, but has been installed as a supplement to the flexible metal conduit (250.118(5)).

So, the secondary conductors are 3 ungrounded phase conductors, a supply-side bonding jumper (neutral) (250.30(A)(2)), and a 3 AWG copper equipment grounding conductor.

The available fault current at the line terminals of the 400 ampere secondary circuit breaker will be:

$$\frac{1.732 \times 6 \text{ feet} \times 20,839.143 \text{ amperes}}{22,965 \text{ "C" value} \times 206.43 \text{ V}} = \frac{216,560.374}{4,740,664.95} = 0.0457$$

$$\frac{1}{1+0.0457} = 0.9563$$

$$\begin{array}{r} 20,839.143 \text{ amperes} \\ \times 0.9563 \\ \hline 19,928.47 \text{ amperes} \end{array}$$

The interrupting rating of the secondary 400 ampere molded-case circuit breaker must be no less than 19,928.47 amperes (110.9).

The arc-flash hazard warning that is required by 110.16(A)(B) and the associated warning label (110.21(B) must be affixed to the 400 ampere circuit breaker enclosure.

The secondary conductors are extended for a length of 75 feet, including the conductors from the transformer secondary to the 400 ampere secondary circuit breaker enclosure and to the downstream 3-phase, 400 ampere enclosed panelboard.

These conductors are installed in steel Electrical Metallic Tubing. There are 4-600 kcmil THHN copper conductors and a 3 AWG THHN copper

104 *Three-Phase Fault Current Analysis*

equipment grounding conductor. According to the information from the Georgia Institute of Technology steal conduit analysis, a 3.5 inch (metric designator 91) steel conduit or tubing with 600 kcmil copper conductors and a 400 ampere overcurrent device and a fault current of 2000 amperes, the EMT length that may be safely used as an equipment grounding conductor with out the need to supplement this raceway with an equipment an equipment grounding conductor is 280 feet. In this example,the EMT is only 75 feet in length (22.86 m). Coupling and connectors used with EMT shall be made up tight (358.42). EMT, factory elbows, associated fittings shall be listed (358.6).

600 kcmil – THHN – 0.8676 square inches

3 AWG – THHN – 0.1562 square inches

$$\frac{\begin{array}{r} 0.8676 \\ \times 4 \end{array}}{3.4704} \text{ square inches}$$

$$\frac{0.0973}{3.5677} \text{ square inches}$$

These conductors require a minimum size 3-1/2 inch EMT (Metric Designator 91) (Table 4 – Chapter 9)-4.618 sq. in.).

The fault clearing time of the 400 ampere-molded and case circuit breaker has been determined to be 0.025 seconds, based on the available fault current of 19,928.47 amperes and the time-current curve characteristics of this circuit breaker. And, the insulation withstand rating of the 600 kcmil THHN copper conductors for 0.025 seconds will be:

$$\frac{600,000 \text{ cm}}{42.25} = 14,201 \text{ amperes } - 5 \text{ seconds}$$

$$14,201 \text{ amperes } \times 14,201 \text{ amperes } \times 5 \text{ seconds } = 1,008,342,005$$

$$\frac{1,008,342,005}{0.025} = 40,333,680,200$$

$$\sqrt{40,333,680,200} = 200,832 \text{ amperes}$$

The insulation withstand rating of the 600 kcmil copper conductors for 0.025 seconds is 200,832 amperes and the available fault current at the 400 ampere secondary circuit breaker is 19,928.47 amperes. So, no problem with 110.10.

Three-Phase Fault Current Analysis   105

The voltage-drop for this feeder circuit, based on a length of 75 feet of conductor length to the 400 ampere panelboard, including the flexible metal conduit from the transformer to the 400 ampere secondary circuit breaker enclosure is as follows:

$$\frac{1.732 \times 12.9 \times 75 \text{ feet} \times 380 \text{ amperes}}{600,000 \text{ cm}} = \frac{636,769.80}{600,000} = 1.06 \text{ volts}$$

$$\frac{\begin{array}{r} 206.43 \\ -1.06 \end{array}}{205.37} \text{ volts line} - \text{to} - \text{line}$$

$$\frac{205.37}{1.732(\sqrt{3})} = 118.57 \text{ volts line} - \text{to} - \text{neutral}$$

The voltage at the 400 ampere panelboard is 205.37 volts line-to-line and 118.57 volts line-to-neutral. This voltage is within the 3% recommended voltage-drop limits of 215.2(A)(2), Informational Note No. 2.

The feeder capacity is 475 amperes @ 90 degree C, with the load on these conductors limited to 380 amperes, due to the nonlinear load. And on the 3-phase, 4-wire wye connected system with the neutral counting as a current carrying conductor (310.15(E)(3)) and the application of Table 310.15(C)(1). The equipment grounding conductor (3 AWG THHN copper) is not counted when applying 310.15(C)(1), (310.15(F)).

The provisions of 110.14(C)(1)(b)(2) apply, as the feeder conductors are THHN (90°C) and the terminal provisions of the 400 ampere main circuit breaker in the enclosed panelboard are based on the 75°C ampacity of the feeder conductors (420 amperes – Table 310.16).

The available fault-current at the line terminals of the 400 ampere, 3-pole molded-case circuit breaker in the 400 ampere panelboard is as follows:

$$\frac{1.732 \times 75 \text{ feet } 19,928.47}{22,965 \text{ "C" Value } \times 206.43 \text{ V}} = \frac{2,588,708.25}{4,740,664.95} = 0.5461$$

$$\frac{1}{1+0.5461} = 0.6468$$

$$\frac{\begin{array}{r} 19,928.47 \quad \text{amperes} \\ \times 0.6468 \end{array}}{12,889.73 \quad \text{amperes}}$$

The available fault-current at the line terminals of the 400 ampere main circuit breaker is 12,889.73 amperes. This establishes the interrupting rating of this

106 Three-Phase Fault Current Analysis

circuit breaker in accordance with 110.9 and, the short-circuit current rating of the panelboard must not be less than 12,889.73 amperes (408.6).

In addition, the interrupting rating of the panelboard circuit breakers must not be less than 12,889.73 amperes (110.9).

The wire bending space in the 400 ampere enclosed panelboard is in accordance with Table 312.6(A), or 8 inches (203mm), based on the 600 kcmil feeder conductors.

The arc-flash hazard warning that is required by 110.16(A)(B) and the associated warning label (110.21(B)) must be affixed to the 600-ampere circuit breaker enclosure.

As we have stated at the beginning of this Chapter, this 400 ampere panelboard is in a separate building, and, as such, the provisions of 250.32 apply. In this case, 250.32(B)(1) and 250.32(B)(2)(a). This building is supplied by a separately derived system (transformer) and there must be an equipment grounding conductor run with the supply conductors and connected to the building or structure disconnecting means (400 ampere panelboard). The 3 AWG copper equipment grounding conductor that is run with the 600 kcmil copper feeder conductors satisfies this requirement. 250.32(A) requires a grounding electrode (system) at this building in accordance with Part III of Article 250, and the 3 AWG copper equipment grounding conductor is required to be connected to the grounding electrode system. This conductor also serves as the bonding means between the grounding electrode at the 150 kVA transformer (250.30(A)(4)) and the grounding electrode (system) at the 400 ampere panelboard (250.50),(250.58),(250.70(A)).

In this case, the grounding electrode system is the metal in-ground support structure bonded to the metal building frame. So, a 1/0 AWG copper grounding electrode conductor will be extended from the neutral busbar in the 400 ampere panelboard to the building steel and a bonding connection is made to the equipment ground busbar in this panelboard, as well (250.68).

A 10 hp-3-phase, 208 volt squirrel-cage, Design B Induction motor is supplied from the 400 ampere panelboard. This 3-phase motor is supplied from a 3-phase, inverse-time circuit breaker. 430.83(A)(2) permits this inverse-time circuit breaker to serve as the controller for this motor. However, there is a separate controller downstream of this circuit breaker and this controller complies with 430.82(A), in that it is capable of starting and stopping the motor and interrupting the locked-rotor current of this 10 hp motor (Table 430.251(A)).

The motor is supplied by THHN-copper conductors in steel electrical metallic tubing and a 6 foot length of flexible metal conduit at the motor

Three-Phase Fault Current Analysis 107

terminal housing. An insulated equipment grounding conductor is run with the branch circuit conductors. The branch circuit length is 60 feet.

The size of the equipment grounding conductors for this motor circuit is based on 250.122(D)(1) and 250.122(A). The size of the motor overcurrent protection may be rated at 250% of the motor full-load current rating from Table 430.52(C)(1) and 430.52(C)(1)(a) permits the use of the next standard size of overcurrent device.

$$\frac{30.8 \text{ amperes}}{\times 2.5}$$
$$77 \text{ amperes}(80 \text{ amperes})(240.6(A))$$

However, we will use a 40 ampere inverse-time circuit breaker and 8 AWG-THHN-copper conductor (40 amperes-60°C). There are no ambient temperature correction factors to be applied in this example from Table 310.15(B)(1). Normally, 240.4 requires conductors to be protected at their ampacity, unless otherwise permitted or required in 240.4(A) thorough (H). Section 240.4(G) permits motor and motor control circuit conductors to be protected in accordance with 430.30 Parts II,III,IV,V,VI,VII. So providing overcurrent protection for the motor circuit conductors in excess of their normal ampacity is permitted by 240.4(G).

The motor branch circuit consists of 3-8 AWG THHN, copper conductors and 1-10 AWG THHN copper equipment grounding conductor (Table 250.122) in a steel EMT raceway and a 6 foot length of flexible metal conduit which connects to the motor terminal housing. Because this is a motor circuit where flexibility is necessary to minimize vibration or to provide flexibility for this equipment (250.118(A)(5)(e)), a wire-type equipment grounding conductor is to be run with the motor branch-circuit conductors. In our example, this conductor is run with the motor circuit conductors.

However, because the electrical metallic tubing is a type of equipment grounding conductor (250.118(4), and the flexible metal conduit does not exceed 6 feet (1.8m) in length, a 8 AWG copper bonding conductor may be installed on the outside of the flexible metal conduit in accordance with 250.102(E)(2).

The size of the electrical metallic tubing from Table 5 and Table 4 of Chapter 9 is 3/4 inch (metric designator 21) for 3-8 AWG-THHN and 1-10 AWG-THHN conductors, or 1/2 inch (metric designator 16) where the equipment bonding conductor is installed on the outside of the flexible metal conduit.

108 *Three-Phase Fault Current Analysis*

And, in accordance with Steel conduit analysis "Vs 1.2" by the Georgia Institute of Technology, the maximum length of the 3/4 inch EMT as the sole equipment grounding conductor for a 40 ampere circuit with a ground-fault current of 500% of the overcurrent device rating (200 amperes) and a 50 volts drop at an arcing fault, the length of the EMT may be 261 feet and serve as the sole equipment of grounding conductor, where the EMT connections are properly tightened (358.52).

3-8 AWG-THHN

1-10 AWG-THHN

Table 5 – Chapter 9

0.0366 square inches–8 AWG-THHN

0.0211 square inches–10 AWG-THHN

$$\frac{0.0366 \text{ square inches}}{0.1098 \text{ square inches}} \times 3 + 0.1098 \text{ square inches}$$

0.0211 square inches

0.1309 square inches

3/4" – 0.213 square inches-metric designator 21 (Table 4–Chapter 9)

The motor will normally be provided with a metal terminal housing (430.12). However, in other than hazardous (classified) locations, substantial nonmetallic housings are permitted.

A means must be provided for the attachment of the equipment grounding conductor at the motor terminal housing in accordance with 250.8(A).

A motor circuit switch is installed "in sight from" (Article 100-Definition-visible and not more than 50 feet (15m)) from the motor. This switch is rated at least 10 hp in accordance with 430.109(A)(1) and the maximum locked-rotor current in amperes for this motor is 179 amperes for the motor disconnect and controller (430.110(C)(1))(Table 430.251(B)).

The voltage-drop on this motor branch-circuit is as follows:

$$\frac{1.732 \times 12.9 \times 60 \text{ feet} \times 30.80 \text{ amperes}}{16,510 \text{ cm}} = \frac{41,289.49}{16,510 \text{ cm}(8 \text{ AWG})} =$$

2.50 volts

Three-Phase Fault Current Analysis 109

205.37 volts (at panelboard)
−2.50 volts (drop)

202.87 volts(at motor)

The phase-to-phase voltage at the 400 ampere enclosed panelboard was determined to be 205.37 volts phase-to-phase and 118.57 volts phase-to-neutral.

Based on a nominal voltage of 208 volts phase-to-phase and 202.87 volts at the motor, the voltage drop equals.

208.00 V

202.87 V

5.13 V

This voltage-drop is within the 5 percent maximum total voltage drop on the feeder and branch circuit supplying this motor load (210.19, Informational Note and 215.2(A)(2), Informational Note No.2).

The available fault current at the 10 hp-3 phase, 208 volt motor is as follows:

$$\frac{1.732 \times 60 \text{ feet} \times 12,889.73 \text{ amperes}}{1557(\text{``C''}) \text{ Value} \times 1 \times 202.87} = \frac{1,339,500.74}{315,868.59} = 4.24$$

$$\frac{1}{1 + 4.24} = 01908$$

12, 889.73 amperes
×0.1908

2459 amperes

The available fault current at the 10 hp 3 phase motor is 2,459 amperes. Due to the rotational energy of the 10 hp motor, the motor fault current contribution is considered to be the motor full-load current rating multiplied by 4.

30.8
×4

123.20 amperes

2, 459.00 amperes
×123.20 amperes

2, 582.20 amperes (available fault-current at motor)

The motor controller includes a 208/120V-500VA control circuit transformer for the motor control circuit.

The motor control circuit conductors are tapped to the load side of the motor branch- circuit, short-circuit and ground-fault protective device,

110 *Three-Phase Fault Current Analysis*

in this case, a 40 ampere, inverse-time circuit breaker. These conductors require overcurrent protection as specified in 430.72(B)(1) or (B)(2), and they will supply the primary of a 500 VA control circuit transformer which is installed within the motor controller enclosure. Table 430.72(B), Column B permits a 14 AWG copper conductor to be protected by a 100 ampere overcurrent device (240.4(G)), and the control circuit transformer primary will be supplied by these conductors and protected by the 40 ampere inverse-time circuit breaker. The terminals must be identified for more than one conductor (110.14(A)) of the specific wire sizes (8 AWG for the motor circuit conductors and 14 AWG for the control circuit conductors. Or, the wire sizes must be changed to accommodate the terminal identification, or another identified connection device must be used.

The control circuit transformer must be protected in accordance with 430.72(C)(1)(2)(3)(4) or (5). In our example, we will use 430.72(C)(1) and the protection will be in accordance with Table 450.3(B). For the transformer primary overcurrent protection, this Table specifies this protection at 167% of the transformer primary current.

$$\frac{500 \text{ VA}}{208 \text{ V}} = 2.40 \text{ amperes} \qquad \frac{\begin{array}{r} 2.40 \text{ amperes} \\ \times 1.67 \end{array}}{4.0 \text{ amperes}}$$

240.6(A) identifies the standard ratings of overcurrent devices, in this case a fuse at 1-3-6-10 and 601A. So, we will use a 3 ampere fuse for the transformer primary protection.

430.72(B), Exception No. 2 recognizes that, due to the fact that this is a two-wire single voltage secondary, the secondary conductors are permitted to be protected by the primary overcurrent device, provided that this protection does not exceed the value determined by multiplying the appropriate maximum rating of the overcurrent device for the secondary conductors from Table 430.72(B) by the secondary-to-primary voltage ratio.

Primary voltage ratio:

$$\frac{\text{secondary}}{\text{primary}} = \frac{120 \text{ volts}}{208 \text{ volts}} = 0.5769$$

45 amperes (Table 430.72 (B)(2), Column C)

$$\frac{\times 0.5769}{2.596 \text{ amperes}}$$

In this example, the transformer protection is a three amperes fuse. So, the secondary conductors require no additional overcurrent protection.

Three-Phase Fault Current Analysis 111

The control circuit conductors are run through 1/2 inch (metric designator 16) steel electrical metallic tubing to the control station (430.73) for a length of 50 feet (15.4 m). And, based on the steel conduit analysis by the Georgia Institute of Technology, the EMT may be installed for a length of 231 feet, where the overcurrent device is 15 amperes, the ground current is 75 amperes, that is, 5 times the rating of the overcurrent device. So, this installation is in compliance with these parameters ,assuming that the connection of the EMT are properly tightened (358.42).

The 120 volt control circuit transformer is required to be grounded in accordance with 250.20(B)(1) and 250.26(1). And, 250.30(A)(5), Exception No. 3 permits a Class 1 control circuit derived from a transformer rated not more than 1000 volt-amperes to be bonded to the transformer enclosure by a bonding jumper sized in accordance with 250.30(A)(1), Exception No. 3 (14 AWG copper or 12 AWG aluminum) where the enclosure is grounded by any of the equipment grounding conductors referenced in 250.118. In our example, the equipment grounding conductor is the electrical metallic tubing (250.118(4) for the branch circuit conductors (250.134(1).

Another important consideration comes from 430.74. Where one conductor of the motor control circuit is grounded, the control circuit is to be arranged so that a ground-fault in this circuit remote from the motor controller will not start the motor and not by-pass manually operated shutdown devices or automatic shutdown devices.

Finally, motor control circuits are required to be on the load side of the motor disconnecting means and arranged so that these circuits are disconnected from all sources of supply when the disconnect is in the open position (430.75(A)). The control circuit transformer installed within the controller enclosure is also required to be connected to the load side of the motor disconnecting means (430.75(B)).

430.72(C)- The motor control circuit transformer must be protected in accordance with 430.72(C). In our example, the control circuit transformer is protected by the 3 ampere primary fuses in accordance with 430.72(C)(2) and Table 450.3(B).

After performing the complete fault-current analysis and sizing the distribution system components accordingly, we cannot overlook the electrical distribution system maintenance. Sizing the system components properly, considering not only the actual connected electrical load, but, in addition the available fault-current, including short-circuit current and the more likely occurrence of ground-fault current, is of vital importance. However, the electrical system maintenance must also be performed to ensure all equipment is working together correctly as a complete system.

112   *Three-Phase Fault Current Analysis*

Overcurrent devices can and do impact arc-flash hazards. Where those devices are poorly maintained, the arc-flash hazards may be markedly increased. Arc-fault circuit interrupters and ground-fault circuit interrupters may not provide the required protection if these devices are not tested in accordance with the manufacturer's instructions.

NFPA 70 E-130.5(B) states that the estimate of the likelihood of occurrence of injury or damage to health and the potential severity of injury or damage to health must take into consideration:

1. The design of the electrical equipment, including its overcurrent protective device and its operating time.
2. The electrical equipment operating condition and condition of maintenance.

There are two informational notes included here as well.

1. Improper or inadequate maintenance can result in increased opening time of the overcurrent protective device, thus increasing the incident energy.
2. For additional direction for performing maintenance on overcurrent protective devices see Chapter 2, Safety-Related Maintenance Requirements.

The key element is that poorly maintained overcurrent protective devices may take longer to clear or possibly not clear at all. This results in longer clearing times or possibly the overcurrent device not clearing at all and an increase in arc-flash incident energies.

Assume that an 800-ampere overcurrent protective device with a six-cycle clearing time (arcing current) is protecting a panelboard (480 V—three phase). However, due to the lack of maintenance, the six-cycle clearing time has been increased to 30 cycles with a significant increase in the arc-flash hazard. NFPA 70E includes several references relating to the maintenance of overcurrent protective devices. These include the following:

205.4 - Requires overcurrent protective devices to be maintained in accordance with manufactures' instructions or industry consensus standards. Maintenance, tests and inspections shall be documented.

210.5 - Requires overcurrent protective devices to be properly maintained so that they will be able to safely withstand or be able to interrupt the available fault current. There is an Informational Note that mentions that lack of or improper maintenance may increase the arc-flash incident energy.

Three-Phase Fault Current Analysis   113

225.1 - This section requires the fuse body and fuse mounting means to be maintained. The mountings for current-limiting fuses should never be altered in any way so that non-current limiting fuses may be inserted in the fuse holders.

Current-limiting fuses have internal parts that do not require maintenance for arc-flash protection. However, the fuse bodies and mountings should be periodically checked and repaired when necessary.

225.2 - This section requires that molded-case circuit breaker handles and cases be properly maintained.

225.3 - Circuit breakers that interrupt fault currents that approach their interrupting ratings must be inspected and tested in accordance with manufacturer's instructions. OSHA 1910.334(b)(2) prohibits reclosing a circuit breaker that has been de-energized by this device until it has determined that this equipment can be safely energized. The repetitive manual re-closing of circuit breakers or reenergizing circuits through replaced fuses is prohibited. If it can be determined from the circuit design that the automatic operation of a device was caused by an overload rather than a fault, no examination of the circuit or connected equipment is needed before the circuit is reenergized.

It must be recognized that when complying with NFPA 70 E-225.3, it is impractical and very likely impossible to determine the level of fault current that is interrupted by a circuit breaker.

NFPA-70B–Recommended Practice for Electrical Equipment Maintenance is an excellent source for information in determining the maintenance frequency as well as the appropriate maintenance procedures. For Example, Chapter 17 of NFPA 70B covers the maintenance of molded-case circuit breakers and Chapter 18 covers fuse and fuse holder maintenance. The NEMA document entitled Guidelines for Inspection and Preventive Maintenance of Molded-Case Circuit Breakers used in Commercial and Industrial Applications is the industry reference for molded-case circuit breaker maintenance.

ANSI/NETA MTS-2011 is the Standard for Maintenance Testing Specifications for Electrical Distribution Equipment and Systems.

- Make a Note that the selective coordination study includes the calculations of arcing currents that are based on the equations of IEEE 1584-2002, the "IEEE Guide for Performing Arc-Flash Hazard Calculations". The total arcing current at each piece of equipment must be determined, as well as the arcing current level that passes through the upstream overcurrent protective device. The arcing current that

114 *Three-Phase Fault Current Analysis*

passes through the upstream overcurrent device is used to determine the clearing time of the overcurrent device.

- To simplify this calculation procedure, Bussmann (Eaton) has a device to calculate available fault current as well as arcing current (855-287-7626). http://toolbox.bussmann.com.

The NEC Sections that apply to selective coordination include the following:

517.31(G) – Overcurrent protective devices serving the essential electrical system shall be coordinated for the period of time that a fault's duration exceeds beyond 0.1 second.

620.62 – Elevators, where more than one driving machine disconnecting means is supplied from the same source, the overcurrent protective devices in each disconnecting means shall be selectively coordinated with any other supply-side overcurrent protective device.

645.27 – Critical Operations Data Systems overcurrent protective devices shall be selectively coordinated with all supply-side overcurrent protective devices.

695.3 – Electric Motor Driven Fire Pumps-Overcurrent protective devices shall be selectively coordinated with all supply-side overcurrent protective devices.

700.32 – Emergency System overcurrent devices shall be selectively coordinated with all supply-side overcurrent protective devices.

701.32 – Legally Required Standby System(s) overcurrent devices shall be selectively coordinated with all supply-side overcurrent protective devices.

708.54 – Critical Operations Power System(s) overcurrent devices shall be selectively coordinated with all supply-side overcurrent protective devices.

Incident Energy and Arc-Flash Boundary

A simple method to apply to determine incident energy (cal/cm2) and the arc-flash Boundary (expressed in inches) when using Bussmann series 1-600A, Low Peak LPS- RK-SP fuses and 601-2000A Low Peak KRP-C-SP fuses is shown in the following Table.

For example, the Arc-Flash Boundary of 6 inches means that anyone performing any work within this distance, including even voltage testing to check that an enclosed panelboard is de-energized, must wear the appropriate personal protective equipment (PPE).

Three-Phase Fault Current Analysis 115

Power System Analysis

Bolted fault current (kA)	1 – 100 A		101 – 200 A		201 – 400 A		401 – 600 A		601 – 800 A		801 – 1,200 A		1,201 – 1,600 A		1,601 – 200 C	
	IE	AFB	IE	AFB	IE	AFB	IE	AFB	IE	AFB	IE	AFB	IE	AFB	IE	AFB
1	2.39	29	>100	>120	>100	>120	>100	>120	>100	>120	>100	>120	>100	>120	>100	>120
2	0.25	6	5.20	49	>100	>120	>100	>120	>100	>120	>100	>120	>100	>120	>100	>120
3	0.25	6	0.93	15	>100	>120	>100	>120	>100	>120	>100	>120	>100	>120	>100	>120
4	0.25	6	025	6	20.60	>120	>100	>120	>100	>120	>100	>120	>100	>120	>100	>120
5	0.25	6	0.25	6	1.54	21	>100	>120	>100	>120	>100	>120	>100	>120	>100	>120
6	0.25	6	0.25	6	0.75	13	>100	>120	>100	>120	>100	>120	>100	>120	>100	>120
8	0.25	6	0.25	6	0.69	12	36.85	>120	>100	>120	>100	>120	>100	>120	>100	>120
10	0.25	6	0.25	6	0.63	12	12.82	90	75.44	>120	>100	>120	>100	>120	>100	>120
12	0.25	6	0.25	6	0.57	11	6.71	58	49.66	>120	73.59	>120	>100	>120	>100	>120
14	0.25	6	0.25	6	0.51	10	0.60	11	23.87	>120	39.87	>120	>100	>120	>100	>120
16	0.25	6	0.25	6	0.45	9	0.59	11	1.94	25	11.14	82	24.95	>120	>100	>120
18	0.25	6	0.25	6	0.39	8	0.48	10	1.82	24	10.76	80	24.57	>120	>100	>120
20	0.25	6	0.25	6	0.33	7	0.38	8	1.70	23	10.37	78	24.20	>120	>100	>120
22	0.25	6	0.25	6	0.27	7	0.28	7	1.58	22	9.98	76	23.83	>120	>100	>120
24	0.25	6	0.25	6	0.25	6	0.25	6	1.46	21	8.88	70	23.45	>120	29.18	>120
26	0.25	6	0.25	6	0.25	6	0.25	6	1.34	19	7.52	63	23.08	>120	28.92	>120
28	0.25	6	0.25	6	0.25	6	0.25	6	1.22	18	6.28	55	22.71	>120	28.67	>120
30	0.25	6	0.25	6	0.25	6	0.25	6	1.10	17	5.16	48	22.34	>120	28.41	>120
32	0.25	6	0.25	6	0.25	6	0.25	6	0.98	16	4.15	42	21.69	>120	28.15	>120
34	0.25	6	0.25	6	0.25	6	0.25	6	0.86	14	3.25	35	18.58	116	27.90	>120
36	0.25	6	0.25	6	0.25	6	0.25	6	0.74	13	2.47	29	15.49	102	27.64	>120
38	0.25	6	0.25	6	0.25	6	0.25	6	0.62	11	1.80	24	12.39	88	27.38	>120
40	0.25	6	0.25	6	0.25	6	0.25	6	0.50	10	1.25	18	9.29	72	27.13	>120
42	0.25	6	0.25	6	0.25	6	0.25	6	0.38	8	0.81	14	6.19	55	26.87	>120
44	0.25	6	0.25	6	0.25	6	0.25	6	0.25	6	0.49	10	3.09	34	26.61	>120
46	0.25	6	0.25	6	0.25	6	0.25	6	0.25	6	0.39	8	2.93	33	26.36	>120
48	0.25	6	0.25	6	0.25	6	0.25	6	0.25	6	0.39	8	2.93	33	26.10	>120
50	0.25	6	0.25	6	0.25	6	0.25	6	0.25	6	0.39	8	2.93	33	25.84	>120
52	0.25	6	0.25	6	0.25	6	0.25	6	0.25	6	0.39	8	2.93	33	25.59	>120
54	0.25	6	0.25	6	0.25	6	0.25	6	0.25	6	0.39	8	2.93	33	25.33	>120
56	0.25	6	0.25	6	0.25	6	0.25	6	0.25	6	0.39	8	2.93	33	25.07	>120
58	0.25	6	0.25	6	0.25	6	0.25	6	0.25	6	0.39	8	2.93	33	24.81	>120
60	0.25	6	0.25	6	0.25	6	0.25	6	0.25	6	0.39	8	2.93	33	24.56	>120
62	0.25	6	0.25	6	0.25	6	0.25	6	0.25	6	0.39	8	2.93	33	24.30	>120
64	0.25	6	0.25	6	0.25	6	0.25	6	0.25	6	0.39	8	2.93	33	24.04	>120
66	0.25	6	0.25	6	0.25	6	0.25	6	0.25	6	0.39	8	2.92	33	23.75	>120
68	0.25	6	0.25	6	0.25	6	0.25	6	0.25	6	0.39	8	2.80	32	22.71	>120
70	0.25	6	0.25	6	0.25	6	0.25	6	0.25	6	0.39	8	2.67	31	21.68	>120
72	0.25	6	0.25	6	0.25	6	0.25	6	0.25	6	0.39	8	2.54	30	20.64	>120
74	0.25	6	0.25	6	0.25	6	0.25	6	0.25	6	0.39	8	2.42	29	19.61	120
76	0.25	6	0.25	6	0.25	6	0.25	6	0.25	6	0.39	8	2.29	28	18.57	116
78	0.25	6	0.25	6	0.25	6	0.25	6	0.25	6	0.39	8	2.17	27	17.54	111
80	0.25	6	0.25	6	0.25	6	0.25	6	0.25	6	0.39	8	2.04	26	16.50	107
82	0.25	6	0.25	6	0.25	6	0.25	6	0.25	6	0.39	8	1.91	25	15.47	102
84	0.25	6	0.25	6	0.25	6	0.25	6	0.25	6	0.39	8	1.79	24	14.43	97
86	0.25	6	0.25	6	0.25	6	0.25	6	0.25	6	0.39	8	1.66	22	13.39	93
88	0.25	6	0.25	6	0.25	6	0.25	6	0.25	6	0.39	8	1.54	21	12.36	88
90	0.25	6	0.25	6	0.25	6	0.25	6	0.25	6	0.39	8	1.41	20	11.32	83
92	0.25	6	0.25	6	0.25	6	0.25	6	0.25	6	0.39	8	1.28	19	10.29	77
94	0.25	6	0.25	6	0.25	6	0.25	6	0.25	6	0.39	8	1.16	18	9.25	72
96	0.25	6	0.25	6	0.25	6	0.25	6	0.25	6	0.39	8	1.03	16	8.22	66
98	0.25	6	0.25	6	0.25	6	0.25	6	0.25	6	0.39	8	0.90	15	7.18	61
100	0.25	6	0.25	6	0.25	6	0.25	6	0.25	6	0.39	8	0.78	13	6.15	55
102	0.25	6	0.25	6	0.25	6	0.25	6	0.25	6	0.39	8	0.65	12	5.11	48
104	0.25	6	0.25	6	0.25	6	0.25	6	0.25	6	0.39	8	0.53	10	4.08	41
106	0.25	6	0.25	6	0.25	6	0.25	6	0.25	6	0.39	8	0.40	9	3.04	34

Arc-flash incident energy

* Bussmann series 1-600 A Low-Peak LPS-RK-SP fuses and 601-2000 A Low-Peak KRP-C-SP fuses, Incident Energy (IE) values expressed in cal/cm2, Arc-flash B (AFB) expressed in inches.

5

Exercise

The following questions are based on references from the National Electrical Code that are directly associated with grounding and bonding. Be sure to review the appropriate code sections, including Exceptions and Informational Notes. This exercise will serve to reinforce your understanding of this important topic. Remember, the NEC is not intended as a design specification (90.2(A)). The scope of this book goes beyond the code references where this is necessary in order to achieve a safe and reliable electrical installation.

As always, I welcome your comments and suggestions.

Gregory P. Bierals

Electrical Design Institute

greg28607@gmail.com

1. Equipment grounding conductors are permitted to be run separately from the PV system conductors within the PV array. However, where the PV system circuit conductors extend beyond the PV array they shall be within the same raceway or cable or otherwise run with the PV array circuit conductors.

    a) True
    b) False

2. Circuit breakers and fuses for PV source circuits shall be _____ for use in PV system dc circuits with the appropriate voltage, current and interrupting ratings.

    a) Identified
    b) Listed
    c) Approved
    d) Recognized

118 *Exercise*

3. The equipment grounding conductors for PV circuits are sized in accordance with _____.

    a) Table 250.122
    b) 250.122
    c) 250.66
    d) Table 250.66

4. A _____ system is permitted to be connected to premises wiring without a grounded conductor if the connected premises wiring or utility system includes a grounded conductor.

    a) Class 1
    b) Communications
    c) Separately derived
    d) Listed Interactive Inverter for a PV system

5. Grounding electrode conductors that are exposed to physical damage shall not be protected in:

    a) Rigid metal conduit
    b) Schedule 40 PVC
    c) Electrical metallic tubing
    d) Cable armor

6. Solidly grounded 3-phase, 4-wire, wye electrical systems supplying promises wiring systems require ground-fault protection of equipment where the system operates at over 150 volts-to-ground and up to 1000 volts phase-to-phase where the service disconnect is rated at _____, or more.

    a) 1200 A
    b) 1000 A
    c) 2000 A
    d) 4000 A

7. Back-fed circuit breakers may be of the plug-in type for use in an enclosed panelboard, but these circuit breakers must be secured by an additional fastener that requires other than a pull to release the device from the mounting means on the panelboard.

    a) True
    d) False

Exercise 119

8. Bonding devices used for the grounding and bonding of the metal frames of PV modules and other equipment must be _____.

   a) Listed
   b) Labeled
   c) Identified
   d) All of the above

9. The DC Ground Fault Detector-Interrupter for PV systems must isolate the _____ conductors, or the inverter charge controller fed by the faulted circuits shall stop the supply to output circuits.

   a) Current-carrying
   b) Equipment grounding conductor
   c) Neutral
   d) Grounded

10. Galvanized steel electrical metallic tubing may be installed in direct contact with the earth.

   a) True
   b) False

11. Grounding electrode conductors in contact with the earth are subject to the minimum cover requirements of 300.5 or 305.15.

   a) True
   b) False

12. Where fences and metal structures are located more than _____ from the substation ground grid they are permitted to have a separate grounding electrode system with the fences and metal structures bonded to this grounding electrode system.

   a) 13 feet (4 m)
   b) 10 feet (3 m)
   c) 16 feet (5 m)
   d) 20 feet (6 m)

13. Surge protective devices shall be connected to ground by a conductor that is not any _____ that necessary and shall avoid _____ bends.

   a) Shorter
   b) Longer
   c) Unnecessary
   d) b and c

120 *Exercise*

14. If rock bottom is encountered and driving a ground rod at up to 45 degrees from the vertical is not possible, the ground rod may be buried horizontally in a trench that is at least _____ deep.

    a) 12 inches
    b) 18 inches
    c) 30 inches
    d) 45 inches

15. A grounding electrode for strike termination devices (lightning protection) must be not less than _____ feet from any other electrode of another grounding system.

    a) 5 feet
    b) 8 feet
    c) 6 feet
    d) 10 feet

16. For optimum paralleling efficiency, multiple ground rods should be spaced at least twice the length of the longest ground rod.

    a) True
    b) False

17. Where auxiliary grounding electrodes are used to supplement the equipment grounding conductors specified in 250.118, the electrode _____ requirements of 250.50, 250.53(C), 250.58, 250.60 and 250.106 do not apply.

    a) Grounding
    b) Bonding
    c) Supplemental
    d) All of these

18. Overcurrent devices for PV dc source circuits shall be readily accessible.

    a) True
    b) False

19. Bonding shall be provided if necessary to ensure electrical continuity and the capacity to safely conduct any _____ current likely to be imposed.

    a) Ground
    b) Fault
    c) Objectionable

Exercise 121

d) Stray

20. Connections between the grounded conductor and the equipment grounding system downstream of the service equipment will produce an _____ current over the equipment grounding conductors.

   a) Fault
   b) Overcurrent
   c) Objectionable
   d) Isolation

21. Nonferrous metal raceway enclosures and cable armor can be used for physical protection of a grounding electrode conductor without bonding at each end.

   a) True
   b) False

22. Where installed inside of buildings, PV system dc circuits exceeding 30 volts or 8 amperes shall be contained in metal raceways and _____ that provides an effective ground-fault current path or in metal enclosures.

   a) AC cable
   b) MC cable
   c) MI cable
   d) Rigid metal conduit

23. PV System DC Circuit Grounding configurations include a two-wire circuit with one _____ grounded conductor.

   a) Resistance
   b) Functionally
   c) Reactance
   d) Solidly

24. A ground ring shall be installed not less than _____ below the surface of the earth.

   a) 18 in.
   b) 42 in.
   c) 60 in.
   d) 30 in.

25. A zigzag connected autotransformer used for the purpose of providing a neutral point for grounding purposes _____ be installed on the load side of any system grounding connection, including those made for the service or for a separately derived system.

122 *Exercise*

a) shall
b) shall not
c) is permitted
d) is not required

26. Two-wire dc systems supplying premises wiring operating at greater than 60 volts and not more than 300 volts shall be _____

a) Isolated
b) Grounded
c) GFCI protected
d) AFCI protected

27. Conductors of ac and dc circuits operating at up to 1000 volts ac or 1500 volts dc are permitted within the same raceway, enclosure or cable, including circuits operating at 480 volts, 240 volts and 120 volts where all of the conductors have insulation voltage ratings equal to at least the maximum circuit voltage applied to any conductor.

a) True
b) False

28. A concrete-encased electrode must have a bare steel reinforcing rod of not less than 1/2 inch in diameter or a continuous 4 AWG bare copper conductor at least _____ feet long.

a) 10
b) 20
c) 30
d) 40

29. Auxiliary or supplemental ground rods connected to metallic equipment frames are permitted to serve as the equipment grounding conductor for the connected equipment.

a) True
b) False

30. Circuit breakers have a minimum interrupting rating of _____ amperes.

a) 10,000
b) 22,000
c) 5,000
d) 1,000

Exercise 123

31. The frame of a portable, trailer mounted, and vehicle mounted generator is not required to be connected to a _____ where the generator supplies equipment mounted on the generator and/or cord-and-plug connected equipment supplied from receptacles mounted on the generator.

    a) Equipment grounding conductor
    b) Grounded conductors
    c) Equipment grounding conductor
    d) Grounding electrode

32. Determine the size of the copper equipment grounding conductor based on its fusing current for a 100 ampere circuit where the ground-fault current has been calculated at 19,240 amperes for 0.025 seconds.

    a) 8 AWG
    b) 6 AWG
    c) 4 AWG
    d) 2 AWG

33. For a ground ring, a copper grounding electrode conductor shall not be required to be larger than _____ unless this conductor extends to the other types of electrodes that require a larger size of conductor.

    a) 6 AWG
    b) 4 AWG
    c) 8 AWG
    d) 2 AWG

34. The requirements of Article 690 pertaining to PV source circuits shall not apply to _____

    a) ac modules or module systems
    b) dc source circuit
    c) Power Source
    d) Array Source

35. The upper limit of the effective voltage level (rms) that a surge-protective device can safely withstand is identified as the _____.

    a) Maximum effective voltage
    b) Maximum continuous operating voltage
    c) Maximum peak voltage
    d) Maximum system voltage

124　*Exercise*

36. The PV maximum source circuit current is calculated by multiplying the sum of the parallel-connected PV module-rated short-circuit currents by 125 percent.

    a) True
    b) False

37. PV source circuits and PV output circuits are not permitted to be contained in the same raceway, cable tray, cable, outlet box, junction box, or similar fitting, with non-PV systems unless the two systems are separated by a partition.

    a) True
    b) False

38. Direct-buried service conductors that are not encased in concrete and that are buried 18 inches or more below grade shall have their location identified by a warning ribbon placed in the trench at least _____ inches above the underground installation.

    a) 6
    b) 10
    c) 12
    d) 18

39. The bonding of a grounding electrode (system) for a lightning protection system to the building grounding electrode system is hazardous and not permitted.

    a) True
    b) False

40. Stationary generators shall be _____

    a) Identified
    b) Listed
    c) Approved
    d) Recognized

41. Switchboards, switchgear and enclosed panelboards shall have a _____.

    a) Circuit directory
    b) Voltage rating
    c) Short-circuit current rating
    d) all of the above

Exercise 125

42. _____ disconnects, unless otherwise marked, shall be considered suitable for backfeed for an interconnected electric power production source.

    a) Identified
    b) Fused
    c) Listed
    d) Field labeled

43. A building or structure supporting a PV system shall utilize a grounding electrode system in accordance with 690.47(B).

    a) True
    b) False

44. PV arrays mounted to buildings are permitted to use the metal structural frame of the building as the grounding electrode (system) if the requirements of _____ are met.

    a) 250.66 (B)
    b) 250.52 (A)
    c) 250.64 (B)
    d) 250.68 (C)(2)

45. Energy storage systems for one-and two-family dwelling units are permitted to operate at _____ volts dc where live parts are not accessible during routine maintenance.

    a) 1000 V
    b) 100 V
    c) 600 V
    d) 1500 V

46. A solar PV system that operates in parallel with and may deliver power to an electrical production and distribution network is known to operate in a(n) "_____ system".

    a) Hybrid mode
    b) Inverted mode
    c) Interactive mode
    d) Internal mode

47. The maximum PV inverter output circuit current is equal to the _____ output current rating.

    a) averege

126　*Exercise*

   b) Peak
   c) Continuous
   d) intermittent

48. Ground-fault protection of equipment is required for 3-phase, 4-wire solidly grounded wye connected systems operating at over _____ to ground and not more than _____ volts phase-to-phase for each service disconnect rated _____ amperes or more.

   a) 150,1000,1000
   b) 480,1000,800
   c) 600,800,1000
   d) 277,600,1000

49. A grounding electrode for the service is not required to be bonded to a grounding electrode for a communications system if the grounding electrode for the communications system is more than 50 feet away.

   a) True
   b) False

50. A high impedance grounded neutral system is recognized as an effective means of reducing _____

   a) Line-to-neutral faults
   b) Line-to-line faults
   c) Neutral-to-ground faults
   d) Arc-flash hazards

ANSWER KEY

  1. a   690.43(C)–250.134(2)
  2. b   690.9(B)
  3. b   690.45
  4. d   200.3, Exception
  5. b   250.64(B)(2)
  6. b   230.95
  7. a   408.36(D)
  8. d   690.43(B)
  9. a   690.41(B)(1)(2)
10. a   358.10(A)(1),(B)(1)
11. b   250.64(B)(4)
12. c   250.194(A)

Exercise 127

13. d 242.24, 250.4(A) Informational Note No. 1
14. c 250.53(A)(4)
15. c 250.53(B)
16. a 250.53(A)(3) Informational Note
17. b 250.54
18. b 690.9(C)
19. b 250.90
20. c 250.6(A),(B)
21. a 250.64(E)(1)
22. b 250.118(10)
23. b 690.41(A)(1)
24. d 250.53(F)
25. b 450.5
26. b 250.162(A)
27. a 300.3(C)(1)
28. b 250.52(A)(3)
29. b 250.54
30. c 250.83(C)
31. d 250.34(A),(B)
32. b 6 AWG copper (22,924.40 A for 0.025 seconds)
33. d 250.52(A)(4), 250.66(C)
34. a 690.6(A)
35. b No NEC reference
36. a 690.8(A)(1)(a)(1)
37. b 690.31(B)(1)
38. c 300.5(D)(3)
39. b 250.60, 250.106
40. b 445.6
41. d 408.4(A), 408.6, 408.58
42. b 705.12(D)
43. a 690.47(A)
44. d 690.47(B)
45. c 706.20(B), Exception
46. c Article 100, Definition "Interactive Mode"
47. b 690.8(A)(1)(c)
48. a 215.10, 230.95, 240.13, 517.17, 700.7(D), 701.7(D), 708.52
49. b 250.50, 250.58, 800.100(A)(B)
50. d 250.36 Informational Note, Annex O-NFPA 70E-2024

128 *Exercise*

To Convert US Custom Units of Measurement to Metric Sizes

Multiply the US measurement by 25.4, or

Divide the metric size by 25.4 to convert to US measurement.

To convert circular mils to metric wire sizes, divide the circular mil area by 1,973.53, or multiply the metric wire size by 1,973.53 to convert to circular mils.

Inch = 0.254 meters
Inch = 2.54 centimeters
Inch = 25.40 millimeters
Meter = 39.37 inches
Millimeter = 0.03937 inch
Centimeters = inches/2.54
Foot = 0.3048 meter
Yard = 0.9144 meters
Mile = 1,609 meters
Kilometer = 0.6213 miles
Square meter = square foot $\div$ 0.093
Circumference = πd
Area of circle = πr^2
Celsius to Fahrenheit = temperature $\times 1.8 + 32$
Fahrenheit to Celsius = temperature -32×0.5556
Square feet to square meter = $m^2 = \dfrac{ft^2}{10.764}$
$\pi = 3.1416$
$\sqrt{2} = 1.414$
$\sqrt{3} = 1.732$

$$\text{Voltage-drop} = \frac{2K \times L \times I}{CM} \text{ - single-phase}$$

$$\text{Voltage-drop} = \frac{1.732K \times L \times I}{CM} \text{ - three-phase}$$

$$\text{Circular mils} = \frac{2K \times L \times I}{VD} \text{ -single-phase}$$

Exercise 129

$$\text{Circular mils} = \frac{1.732K \times L \times I}{VD} \text{ - three-phase}$$

$K = 12.90$ ohms $-$ copper

$K = 21.20$ ohms $-$ aluminum

$L =$ one way length in feet of conductor

$I =$ amperes of load

$CM =$ circular mil area of conductor (NEC Table 8-Chapter 9)

Neutral current in a three-phase, four-wire wye-connected system
$$\sqrt{L1^2 + L2^2 + L3^2 - [(L1 \times L2) - (L2 \times L3) - (L1 \times L3)]}$$

Series Circuits

Total resistance, $R_T = R_1 + R_2 + R_3 + R_4$
   Resistance where voltage and power (wattage) are known

$$R = E^2 P$$

Power (wattage), $P = I^2 \times R$
Voltage equals the sum of all of the power supplies

$$V_T = V_1 + V_2 + V_3 + V_4$$

Current equals, $I = E$ (Source) $/R$ (Total)

Parallel Circuits

The voltage – drop across each resistance is equal to the voltage of the power source.

   The current in each branch of the parallel circuit is calculated by the formula: $I = \frac{E}{R}$

   Parallel Circuit Power (Wattage) in each branch equals $P = I^2 R$ or $P = E \times I$ or $P = \frac{E^2}{R}$

   Total Power – The sum of the power in each branch resistance is always less than the smallest resistance.

   If all of the resistors have the same resistance, the total resistance equals the resistance of one resistor, divided by the total number of resistors in parallel.

130 *Exercise*

Example

What is the total resistance of 5 – 10 ohm resistors in parallel?

$$\frac{10 \text{ ohms}}{5 \text{ resistors}} = 2 \text{ ohms Total}$$

Where the paralleled resistors have different ohmic ratings, the total resistance may be calculated using the <u>Product Over Sum Method</u>

Example

Two resistors in parallel −5 ohms − 7 ohms

$$\begin{array}{cc} 5 \text{ ohms} & 5 \text{ ohms} \\ \times 7 \text{ ohms} & +7 \text{ ohms} \\ \hline 35 \text{ ohms} & 12 \text{ ohms} \end{array}$$

$$\frac{35 \text{ ohms}}{12 \text{ ohms}} = 2.92 \text{ ohms}$$

Total resistance = 2.92 ohms

If a third resistor is added (fourth, fifth, sixth, etc.), the same formula may be used to calculate the total resistance.

Example

Two resistors are added, one is 10 ohms and the second is 15 ohms, what is the total resistance?

$$\begin{array}{cc} 2.92 \text{ ohms} & 2.92 \text{ ohms} \\ \times 10 \text{ ohms} & +10 \text{ ohms} \\ 29.2 \text{ ohms} & 12.92 \text{ ohms} \end{array}$$

$$\frac{29.2 \text{ ohms}}{12.92 \text{ ohms}} = 2.26 \text{ ohms}$$

$$\begin{array}{cc} 2.26 \text{ ohms} & 2.26 \text{ ohms} \\ \times 15 \text{ ohms} & +15 \text{ ohms} \\ 33.9 \text{ ohms} & 17.26 \text{ ohms} \end{array}$$

$$\frac{33.9 \text{ ohms}}{17.26 \text{ ohms}} = 1.96 \text{ ohms}$$

Exercise 131

Total resistance $= 1.96$ ohms

The reciprocal method may be used to calculate the total resistance where the resistances have different ratings.

Example

Four resistors in parallel with ratings of 10 ohms $-$ 6 ohms $-$ 8 ohms $-$ 15 ohms

$$\frac{1.00}{\frac{1}{10} + \frac{1}{6} + \frac{1}{8} + \frac{1}{15}}$$

$$\frac{1}{10} = 0.1 - \frac{1}{6} = 0.167 - \frac{1}{8} = 0.125 - \frac{1}{15} = 0.067$$

$$
\begin{array}{r}
0.1 \\
+\,0.167 \\
+\,0.125 \\
+0.067 \\
\hline
0.459
\end{array}
\qquad
\frac{1.00}{0.459} = 2.178 \text{ ohms}
$$

Total resistance equals 2.178 ohms.

Index

A

Answer Key 126

B

Back-Up Generators 51
Bonded 21, 22, 28, 46, 51, 102, 119
Bonding Conductor 13, 28, 50, 98, 102
Bonding Jumper, Main 29
Bonding Jumper, Supply-Side 30
Bonding Jumper, System 30

E

Electrical System Maintenance 111

F

Fault-Current 2, 29, 80, 83, 105, 111
Fault-Current, Available 2
Fault Protective Device 109

G

Ground 1, 11, 12, 45, 77, 119
Grounded Conductor 2, 13, 25, 30, 90, 118
Grounded, Functionally 26
Grounded, Solidly 12
Ground-Fault 2, 11, 39, 68, 89, 111, 118, 126
Ground-Fault Circuit Interrupter 12, 14

Ground-Fault Circuit Interrupter, Special Purpose 12
Ground-Fault Current Path 2, 9, 14, 70, 94, 121
Ground-Fault Current Path, Effective 2
Ground-Fault Protection of Equipment 15, 91, 93, 118, 126
Grounding Electrode 13, 15, 17, 25, 49, 51, 84, 119, 121, 126
Grounding Electrode Conductor 13, 20, 25, 53, 95, 121, 123

I

Inverter, Stand Alone 27

M

Microgrid 27

O

Overvoltage Protection 52

P

Panelboard, Enclosed 27
PV (Photovoltaic) System 28
PV System-Grounding and Bonding 40

S

Service Conductors 13, 28, 56, 67, 77, 124

134   *Index*

Service Entrance Conductors 26, 28, 77

Single-Phase Fault Current Analysis 55, 66

Surge Protection 42, 44

T

Table of Conductor Fusing Currents 90

Three-Phase Fault Current Analysis 77

About the Author

Gregory P. Bierals is the president and Technical Director of Electrical Design Institute. He has presented technical seminar programs on the topics of the national electrical code, designing overcurrent protection, grounding electrical distribution systems, electrical systems in hazardous (classified) locations, and grounding solar photovoltaic systems. These courses have been offered throughout the United States and abroad since 1978. His published books include The NEC and You, Perfect Together, Grounding Electrical Distribution Systems, Designing Overcurrent Protection, NEC Article 240 and Beyond and Grounding and Bonding Photovoltaic and Energy Storage Systems.

About the Author

Gregory P. Bierals is the president and technical Director of Electrical Design Institute. He has presented technical seminar programs on the topics of the national electrical code, designing overcurrent protection, grounding electrical distribution systems, electrical systems in hazardous (classified) locations, and grounding solar photovoltaic systems. These courses have been offered throughout the United States and abroad since 1978. His published books include The NEC and You, Perfect Together, Grounding Electrical Distribution Systems, Designing Overcurrent Protection, NEC Article 240 and Beyond, and Grounding and Bonding Photovoltaic and Energy Storage Systems.